Fearing Sellafield

What it is and why the Irish want it shut

Colum Kenny

Gill & Macmillan

Gill & Macmillan Ltd
Hume Avenue, Park West, Dublin 12
with associated companies throughout the world
www.gillmacmillan.ie

© Colum Kenny 2003
0 7171 3583 7

Index compiled by Cover To Cover
Print origination by Red Barn Publishing
Printed by ColourBooks Ltd, Dublin

This book is typeset in Berkley Book 12 on 14.5pt.

*The paper used in this book comes from the wood pulp of managed forests. For every
tree felled, at least one tree is planted, thereby renewing natural resources.*

A CIP catalogue record for this book is available from the British Library.

1 3 5 4 2

for
Oisín, Conor and Sam
and their generation

The principal function of the Sellafield site is the reprocessing of irradiated uranium fuel for the recovery of uranium, the separation of plutonium, and the storage of the highly active fission products which form the main radioactive waste material in the irradiated fuel.

The United Kingdom Health and Safety Executive, 1983.

The safety of the storage of highly active liquor (HAL) has consistently been one of the more significant public concerns associated with the reprocessing of irradiated nuclear fuel and the management of nuclear waste. This was evident at the Windscale Inquiry of 1977 and during the commissioning of THORP in 1993 and 1994.

The United Kingdom Nuclear Installations Inspectorate, 2000.

A successful sabotage attack on a nuclear facility could cause widespread radioactive contamination and loss of life.

The United Kingdom Director of Civil Nuclear Security, 2002.

Contents

Acknowledgements

Thanks to British Nuclear Fuels for an interesting visit to Sellafield, and for further information. The Royal Irish Academy and Friends of the Earth, as well as various British and Irish government departments and official agencies mentioned in the text below, were also helpful. So, too, were Nuala Ahern, MEP, and Veronica McDermott of PAN Ltd, which acts for BNFL in Ireland.

I wish to express my appreciation of the support of Dublin City University, which encourages its academic staff to publish works of cultural and social value to the community at large. I am grateful to Fergal Tobin of Gill & Macmillan for his enthusiasm and advice when I first put the idea of this book to him. And to Catherine Curran. Thanks for the coffee and ideas, Catherine.

Abbreviations and Websites

BE   British Energy. Privately owned nuclear power company.
              www.british-energy.com
BNFL   British Nuclear Fuels Limited. State-owned. Runs Sellafield.
              www.bnfl.com
COMARE   Committee on Medical Aspects of Radiation in the
              Environment (UK). www.doh.gov.uk/comare/comare.htm
CORE   Cumbrians Opposed to a Radioactive Environment.
              www.corecumbria.co.uk
DOELG   Department of the Environment and Local Government
              (Ireland). www.environ.ie/main.html, with a useful link for
              nuclear safety information at
              www.irlgov.ie/tec/energy/nuclear
DEFRA   Department for Environment, Food and Rural Affairs (UK).
              www.defra.gov.uk/environment, with links to 'radioactivity'.
DTI   Department of Trade and Industry (UK).
              www.dti.gov.uk/nid/index.htm, for nuclear industries
              regulation.
EA   Environment Agency (England and Wales).
              www.environment-agency.gov.uk
EHSNI   Environmental and Heritage Service of the Department of
              the Environment (Northern Ireland). www.ehsni.gov.uk
EPA   Environmental Protection Agency (Ireland). www.epa.ie
EU   European Union. Useful resources are http://europa.eu.int/
              comm/environment/radprot/index.htm#studies
              and
              http://europa.eu.int/comm/energy/nuclear/index_en.html
Euratom   European Atomic Energy Community Treaty. Ensures the
              establishment of the basic installations necessary for
              developing nuclear energy, as well as the regular and
              equitable supply of ores and nuclear fuels among Member
              States. The latter is achieved, under the supervision of the
              EU Commission, by the Euratom Supply Agency.
              www.europa.eu.int/comm/euratom/index_en.html
FOE   Friends of the Earth. www.foe.co.uk for England, Wales and
              Northern Ireland; www.foe-scotland.org.uk;
              www.iol.ie/~foeeire for the Republic of Ireland.
GI   Greenpeace International. www.greenpeace.org

HSE	Health and Safety Executive (UK). www.hse.gov.uk, including at 'nuclear' under A–Z index.
IAEA	International Atomic Energy Agency. www.iaea.org
INES	International Nuclear Event Scale. For rating accidents.
ITLOS	International Tribunal for the Law of the Sea. Established under UNCLOS (below). www.itlos.org, which includes written and oral proceedings of the Sellafield MOX plant case (Ireland v. UK).
LMA	Liabilities Management Authority. No website yet.
NEA	Nuclear Energy Agency of the Organisation for Economic Co-operation and Development (OECD). www.nea.fr
NII	Nuclear Installations Inspectorate of the HSE/NSD. See NSD website for information on its work.
Nirex	Nuclear industry group examining safety, environmental and economic aspects of the deep geological disposal of radioactive waste. www.nirex.co.uk
NRPB	National Radiological Protection Board (UK). www.nrpb.org.
NSD	Nuclear Safety Directorate (UK). www.hse.gov.uk/nsd/index.htm
OCNS	Office for Civil Nuclear Security (UK). No website, but some information at DTI.
OECD	See NEA above.
OSPAR	Convention (of the Commission) for the Protection of the Marine Environment of the North-East Atlantic. Known as 'OSPAR' because it has been ratified by the contracting parties to the older Oslo or Paris Conventions. www.ospar.org
PIU	Performance and Innovation Unit (UK Cabinet Office). www.piu.gov.uk
RPII	Radiological Protection Institute of Ireland. www.rpii.ie
SEPA	Scottish Environment Protection Agency. www.sepa.org.uk
SLLC	Sellafield Local Liaison Committee. www.sllc.co.uk
STOA	Scientific and Technological Option Assessment Programme of the European Parliament. It commissioned the 'WISE' report (see WISE). www.europarl.eu.int/stoa
WISE	World Information Service on Energy, Paris. Authors of a controversial report for STOA on the effects of reprocessing at Sellafield and Cap de la Hague. www.wise-paris.org
UKAEA	United Kingdom Atomic Energy Authority. www.ukaea.org.uk
UNCLOS	United Nations Convention on the Law of the Sea. Led to establishment of ITLOS (see ITLOS).
Urenco	Uranium enrichment company established by BNFL with Dutch and German partners. www.urenco.com

CHAPTER 1

What is Sellafield?

Fears about Sellafield have grown since the unprecedented terrorist attacks in the United States on 11 September 2001. Images of jet airliners plunging into the Twin Towers of Manhattan aroused alarm on both sides of the Irish Sea. How vulnerable is Sellafield to a similar attack, or to some other disaster? How serious would be the consequences of such an event? People were already asking what might happen if a ship carrying radioactive fuel to, or from, Sellafield were to sink near land; or were deadly liquids stored at the nuclear plant to leak. Now, they have more urgent questions. For, even if debates surrounding the 2003 British White Paper on Energy end with a decision not to permit the building of new nuclear power stations, Sellafield itself remains operational and dangerous.

A disaster at Sellafield would threaten both Britain and Ireland. Located just south of the Scottish border in Cumbria, England, the plant is little more than one hundred miles from the Irish coast. Any plume of radioactivity released from 'THORP' or 'MOX' or the other nuclear facilities of Sellafield could soon be borne on the wind to Glasgow or Edinburgh, Leeds or Manchester, Cardiff or Dublin, and onwards across these islands to the wider world. A disaster is likely to result in many deaths.

Sellafield reminds us of our mortality. We would prefer that it did not threaten us, but we feel powerless to do much in the shadow of its looming presence and in the wake of passing ships that transport its deadly products. However, during the spring of 2002, my fatalism about its continued existence was transformed into a determination to discover more about Britain's nuclear industry and its activities. This book is one result of that decision.

It was my sons who alerted me to the extent of my complacency about Sellafield. Sam came home from school one day deeply worried by what he had heard during a class discussion on nuclear power. He said that he would never again feel the same about the world. Then aged twelve, he immediately typed and posted an eloquent appeal to Tony Blair, asking the prime minister to 'please close Sellafield' in order to prevent cancers and to avoid pollution that 'will pass over you and kill your plants, animals and people'. Sam received no reply from Downing Street.

My determination was stiffened by a remark made by Sam's brother, Conor. On a fine summer evening, he returned from swimming with his friends at 'The Forty Foot' rock-pool in Dublin Bay and asked me if I knew that 'the Irish Sea is the most radioactive sea in the world'. It is a description that one hears from time to time, dropped almost casually. But hearing it out of the mouth of my son made me uncomfortable. Is it true and, if so, what does that fact mean for our welfare? Are we resigned to accepting the discharge of radioactive wastes from Sellafield into open waters in which we and our children bathe?

I have dedicated this work to my sons, not as a sentimental tribute, but because they and future generations must live with the consequences of what their parents and grandparents have done with the atom and with the industries that exploit our knowledge of its power. Deadly wastes produced by the creation of electricity from nuclear energy will not go away.

But first things first. What is Sellafield? It is *not* a big power plant. Contrary to one popular assumption, it does not generate vast amounts of electricity for the national grid of the United Kingdom. Electricity has certainly been produced at its Calder Hall reactors, the four towers of which are among Sellafield's most visually distinctive features. However, that electricity has been primarily and principally for the needs of the Sellafield plant itself.

The actual, chief purpose of Sellafield is the recycling of old, spent nuclear fuel from power plants in Britain and abroad. This reprocessing involves the manufacture of new fuel, most notably today MOX (Mixed Oxide) fuel, which contains plutonium in addition to enriched uranium. Reprocessing is both very expensive and highly lucrative. A secondary but significant function of Sellafield is to act also, in practice, as a store for highly dangerous nuclear wastes. The absence of adequate long-term storage arrangements for nuclear waste in Britain, as well as the failure of British Nuclear Fuels (BNFL) to return to customers certain by-products of reprocessing that were intended to be returned to them, means that Sellafield's secondary function as a store has become far more significant and far more hazardous than was earlier meant to be the case.

In particular, there is certain highly radioactive liquid waste that has long been kept in tanks at Sellafield and that remains extremely dangerous. Were this to escape in any form, its release could result in very many more deaths than did the release of radioactivity following the disaster at the power plant in Chernobyl.

On both sides of the Irish Sea, people are deeply concerned about the management of reprocessing and storage facilities at Sellafield. They worry too about the transportation of nuclear fuels to and from the plant, and about the authorised and unauthorised release by BNFL of certain radioactive materials into the air and sea.

The Irish government has initiated a number of unprecedented international legal actions against Sellafield. Parliamentarians in Britain are becoming more aware of the reasons for Ireland's concerns through the work of the British-Irish Inter-Parliamentary Body and of the British-Irish

Council. The council is one of the fruits of the Good Friday Agreement on Northern Ireland. Representatives from Ireland and from the Isle of Man, which lies near the coast of Cumbria where Sellafield stands, have ensured that the reprocessing plant will be discussed at its meetings. In Scotland, too, questions are raised in the Scottish Parliament about the nuclear facility, although ministers appear reluctant to discuss a matter in relation to which they have limited powers. A number of Scottish and Welsh MPs have also voiced their concerns at Westminster. On 22 November 2002, in advance of a British-Irish Council meeting at New Lanark, the Scottish National Party's Shadow Environment Minister, Mr Bruce Crawford MSP, said,

> The meeting of the BIC today is an excellent opportunity for Scotland to stand shoulder to shoulder with Ireland in a bid to reverse the UK government's decision to have Mixed Oxide Fuel (MOX) production at Sellafield in England ... Clearly the Irish treat the threat of nuclear pollution coming from a normally friendly neighbour very seriously ...
> It is important for a small nation like Scotland to stand by Ireland on such a vital environmental issue.

I visit Scotland from time to time, usually to spend a few days in the beautiful valley of Eskdalemuir in eastern Dumfriesshire. From there, the River Esk flows down, across the English border, and through Cumbria into the sea, very close to Sellafield. Neither rivers nor radiation are respecters of national boundaries and the consequences of a disaster at Sellafield could be catastrophic for Scotland's largest centres of population. However, most of Scotland's electricity is nuclear in origin and there has been a certain political reticence on the part of some Scots to acknowledge those possible consequences. I hope that this book helps to raise awareness of the issues that are important in relation to Sellafield, in both Britain and Ireland.

The attitude of successive Irish governments towards

Sellafield has been hostile but also, until recently, quite restrained. Fianna Fáil, the dominant party in the current coalition government, broke an explicit electoral promise to fund fully a private legal action against Sellafield by residents of Louth, which is the county of the Republic of Ireland that is closest to Cumbria. And, for many years, Irish governments assigned effective responsibility for nuclear matters to junior ministers.

It must also be said that, in the 1970s, the Irish government itself was on the point of building a nuclear power station, and that its own record on the environment is not without blemish. Indeed, it may yet happen that the Republic of Ireland will run short of sufficient electricity to fuel its current economic growth and will need to buy power from the United Kingdom. In such a case, it may prove impossible or impractical to distinguish between the origins of that power in order to screen out energy that has reached the British national grid from nuclear sources. However, to the extent that the Irish government and others are now helping to ensure that the British nuclear industry operates as safely as possible, and that its wastes are stored in an appropriate manner, their efforts are widely welcomed by many people.

The nuclear industry is at a turning point. In 2003, UK government ministers are reported to be divided about its future. Recent developments suggest that many British and European policy makers favour its continued support and expansion, rather than the contraction and closure advocated by environmentalists and others. Indeed, supporters of nuclear power now claim that it is the best environmental option, being a cleaner method of generating electricity than are methods using either gas or coal. In the 2002 annual report of British Nuclear Fuels (BNFL), which owns and operates Sellafield, that company's chairman warned that, 'If the [UK] government does not set the policy framework to allow new stations to be commercially built, then nuclear generation will not be an option for the future. We will have missed the boat, with severe consequences for the environment and the long-term security of the UK's energy supply.'

The chairman, Hugh Collum, admitted that, 'Linked to any decisions on the future role of nuclear energy is the issue of long-term radioactive waste storage.' It is precisely the dangers posed by that waste and by other radioactive materials, especially in the event of any successful sabotage attack on a nuclear facility, that makes the cost of running Sellafield, in particular, too high for its opponents. In the worst envisaged circumstances, a disaster at Sellafield could become a catastrophe for Ireland or Britain, resulting possibly in hundreds of thousands of deaths and the lasting pollution of cities and countryside. A scientific study by Taylor, published in 1994 for a number of British local authorities, calculated the incidence of fatalities and cancers, among other damage, likely to result in the specific areas of Liverpool, Manchester, South Yorkshire, Leeds and Bradford, Glasgow and Strathclyde as a consequence of a catastrophic failure of the 'Highly Active Liquid Waste' tanks at Sellafield. Using complex software provided by the European Commission, Taylor painted a gloomy picture of many deaths, great economic impact and the long-term relocation of large populations.

A catastrophe may never occur, and probably will not, but opponents of nuclear energy argue that the likely consequences of such an event would be so awful that they merit the closure of Sellafield as soon as possible, as well as the removal of its accumulated radioactive waste to a safe underground storage facility.

The nuclear industry itself is complex, and the issues relating to nuclear energy are not simple. It is my intention to guide the reader through these as clearly and fairly as possible, in the hope that people may become better equipped to respond to the reality of Sellafield than I was when confronted by my sons' concerns about it.

CHAPTER 2

Visiting Sellafield

A black limousine speeds up the motorway. I have been met at Manchester Airport by a neatly dressed driver, holding a sign that reads 'BNFL Sellafield'. Now, in his company's limo Saab in the fast lane of the M6, we are bound for Britain's most controversial nuclear plant, by way of the port of Barrow-in-Furness. Out of Barrow, ships ply their trade through the Irish Sea, carrying radioactive cargoes. These ships are owned by a subsidiary of British Nuclear Fuels (BNFL), called Pacific Nuclear Transport Ltd (PNTL). The vessels are met at Barrow by special trains that transport great 'flasks' of radioactive fuel to and from the reprocessing facilities at Sellafield itself.

My companions in the long, sleek car are two members of a television crew and a reporter from a Dublin radio station. We are guests of BNFL, coming to see Barrow and Sellafield with our own eyes. BNFL is happy to meet all of our costs, which include Aer Lingus return flights from Dublin to Manchester at €296 each. Such trips are frequently organised for journalists and other favoured guests. They are a privilege, especially since access to the site became even more restricted after the events of 11 September 2001. The general public must make do with a Visitors Centre, which stands outside the gates of Sellafield and which has recently been refurbished. It is uncertain in what way, if any, BNFL's generous hospitality later influences media reports or policy decisions.

Barrow and Sellafield lie on the north-west coast of Cumbria, an area that industrialists and tourists have tended to avoid. This is out beyond the popular Lake District, the peaks and ponds of which are famously celebrated in the poetry of William Wordsworth.

As we enter Barrow, we find ourselves driving through the streets of a town that has seen better days. Once a busy port, exporting iron ore and supporting bigger shipbuilding yards than it now does, it is hungry for work. The terminal of PNTL lies in a quiet part of the harbour, and is certainly no hive of activity on the day that we arrive. PNTL runs seven ships. Just two are in port for our inspection, and neither is currently loaded with nuclear cargo. But, heading for Barrow as we visit are two vessels bringing back a controversial shipment of MOX fuel from dissatisfied customers in Japan.

There is no outward sign of high security at the terminal gates. Heavily armed guards are needed only when radio-active fuels such as plutonium are present. The standard security seems no greater than what one might expect at any industrial facility. However, there is a warning notice posted. We should beware that we were entering a 'multi-hazardous area' and that CCTV cameras are in operation.

We are met at Barrow by Captain Malcolm M. Miller, Head of Operations Transport at BNFL, and by Alan Hughes, Media Affairs Manager for the company. Now, as later, we are treated with great courtesy by the representatives of BNFL. But we are also bound by a tight schedule, which leaves no time for meeting anyone other than BNFL personnel. Miller, a Glaswegian with a fine Scottish accent, spent many years on the high seas before becoming desk-bound. We embark with him on a tour of the *Pacific Crane*, which will be described in Chapter 8. The tour is intended to rebut criticisms that the transportation of nuclear fuel by sea is intrinsically hazardous. Following that tour, we take tea and cakes with Captain Miller, sitting in a meeting room, beneath a portrait of Queen Elizabeth II. He is polite and informative, but never lets his guard down when answering questions.

However, we are eager to get to Sellafield. From Barrow, our journey, by limousine, continues along winding country roads. Crawling behind tractors, we gaze out on farms recently threatened with obliteration by foot-and-mouth disease. The terrain around Sellafield is old agriculture country, and few areas of England are more remote.

Arriving at Sellafield, we are brought directly to stay the night at Sellapark House, a private hotel run by BNFL. It would be fair also to call the place 'Sellafield View'. For, from the back window of what other hotel on earth can one gaze across a field at a massive nuclear reprocessing plant? Sellapark House has fishing rights to a quarter mile stretch of the River Calder, 'providing excellent salmon and sea trout fishing within its own grounds'. One suspects that marketing 'Sellafield Salmon' and 'Sellafield Sea Trout' might present quite a challenge.

The décor of this ancient red-brick house is sumptuous. There are deep armchairs, a dining room appointed in leather and wood (containing another portrait of Queen Elizabeth), and thickly carpeted bedrooms. A free bar is well stocked, with dozens of different whiskeys and a special local beer. 'At a glance', proclaims the glossy, coloured brochure which guests at the BNFL hotel receive, 'Sellapark gives the appearance of a seventeenth century house with nineteenth century additions. Closer examination, however, reveals earlier features such as window embrasures of the fifteenth or sixteenth century and the remains of a fourteenth century type window recess and seat.'

Sellapark is a hotel where you never get the bill. Wine flows generously and our meal starts with prawns. Are they making a point about how safe the local shellfish is? The choice of main course is either calf's liver or sea bass, and I am torn between thoughts of cattle diseases and nuclear outflows. The dessert of strawberries on a bed of mousse is delicious.

But dinner is also work. It is hosted by Dr Rex Strong, a physicist and the Head of Site Environmental Management at

Sellafield. Also present are three of BNFL's media relations people. As the evening wears on, the opportunity is taken by BNFL representatives to express disagreement with Greenpeace, and to comment gently upon the efforts of 'Bono [of U2] and the fragrant Ali' (his wife) who have argued publicly against nuclear energy. Less gentle is the judgement of Dr Strong on the decision of the Irish government to circulate iodine tablets, which, as we shall see again later, he describes as 'misleading' and 'a deception'.

We are scheduled to be collected from the BNFL hotel at 8.15 the following morning, to be taken on a drive through Sellafield and on a tour of the THORP (Thermal Oxide Reprocessing Plant) and MOX facilities. Sellafield is about two kilometres long and two-and-a-half kilometres wide, stretching over 600 acres and including many buildings.

Rising early, I decide to take an unplanned walk by myself up to the main gate of the Sellafield site. By 7.30 a.m., on the narrow approach road, there is a two-mile tailback of crawling traffic. More than 10,000 people work at Sellafield and it is clearly time for a change of shift. Turning into another road that runs along the high perimeter fence, I take a couple of photographs of Sellafield through the wire. A uniformed member of the United Kingdom Atomic Energy Authority Constabulary is despatched to intercept me on my way back to the hotel. Had a passing motorist called, or was I being observed on camera? He states that I have taken some photographs but seems reassured when I tell him that I am with a group scheduled to visit the plant later that day. He says with a chuckle that such photographs as I had taken are not the subject of a D[A]-notice (official 'defence advisory' censorship notices issued to the UK media concerning certain highly classified matters of security).

Returning to the BNFL hotel, I find another black limousine waiting to take us on the short drive to Pelham House, a magnificent country mansion owned by BNFL and lying just outside the perimeter of Sellafield. Here, in an administrative office, we are shown a corporate video and given a briefing to prepare us for what to expect inside the facilities that we are about to visit. We are told that we may not carry mobile phones, chew gum or take snuff, among other prohibitions. And my notebook will need to be monitored as I leave, but only for radioactivity as I am pleased to learn.

Two later chapters will be devoted to describing THORP and MOX. Both facilities are remarkable, not simply because of the presence of plutonium, but because the insignificant appearance of that deadly substance contrasts so starkly with the awesome sight of the massive pond in which flasks of spent nuclear fuel are stored. Vast sums of money have been spent building THORP and MOX, and the activities within them are crucially important in determining the policy of the United Kingdom government on nuclear power.

Before we enter THORP and MOX, we are driven through other parts of the Sellafield site and shown the stumps of the two old Windscale stacks. It was here that fire broke out in 1957, and from here that dangerous radioactive material escaped into the atmosphere. A request by me to go inside the facility that is designated 'B215', where highly active liquid waste is stored, is turned down on the grounds that I had not given sufficient notice.

Later, as our visit ends and we depart from Sellafield through a back gate, we pass twice over that innocuous-looking pipeline that has, for years, discharged radioactive wastes into the Irish Sea, which borders the site. The pipeline is pointed out to us without comment.

On leaving Sellafield, as upon entering it, our driver has to negotiate his way around a makeshift chicane of concrete lengths thrown up in the aftermath of the attacks in the United States on 11 September. I notice that a special guard, who stands nearby, discreetly nestles an automatic weapon.

And so, it is again time for some BNFL food. A pleasant buffet lunch is hosted at Pelham House by the manager of the MOX plant. When it ends, we are whisked away to Manchester Airport on a journey that takes about three hours. As our limousine glides down the M6, at high speed, its passengers drift into sleep. It is surprisingly hard work just visiting Sellafield.

We had gathered at Dublin Airport at 8.50 a.m. on Monday and arrive back in Dublin at 8.00 p.m. Tuesday. The trip has been very demanding, with six hours on the road, and working lunches and dinner, and long walks through plants in close proximity to uranium and plutonium. We have sought and received a continuous flow of complex information, and have tried to absorb it all. Repeated security checks, changes of protective clothing and radiation monitoring have also taken their toll.

After receiving so much assistance and hospitality, I feel almost offensive when I inform BNFL representatives that I wish to reimburse them for the cost of my airline ticket. It would be impractical to propose also making a donation towards the petrol for their company cars, and possibly sheer bad manners to ask to pay for the food and drink that I have consumed as their guest.

BNFL is not unique in its generosity to special visitors. Corporate hospitality is a common phenomenon. It helps to create an environment in which journalists, politicians and others may be more receptive than otherwise to the messages of a host corporation. Psychologically, it is more difficult to be harsh to those whom we have come to know and who have treated us agreeably than to those whom we have not met personally and to whom we do not feel somewhat obliged. I had requested permission to visit Sellafield and permission was granted. Others are invited without first expressing an interest. In any event, it is reasonable for BNFL to treat its visitors well and to explain its processes and put its points of view. However, it is also reasonable to assume that visitors are influenced by those explanations and views.

Local organisations such as CORE (below) have not the resources to match the level of hospitality provided by BNFL. Phoning such groups, visiting their websites or getting them to mail material is a less pleasurable and, it may be assumed,

less influential experience than a visit to Sellafield as a guest
of BNFL.

Core Values
Sellafield stands just north of the small town of Seascale, on
the coast of Cumbria. Cumbria, in general, is a beautiful
place and is thinly populated. Its boundaries are the Irish Sea
to the west (from the Solway Firth to Morecambe Bay), the
Scottish border to the north and the Pennine hills to the east.
It is physically dominated by the Lake District, which is now
a national park, and its biggest town is Carlisle.

Many people in Cumbria support the Sellafield plant and
many Cumbrians benefit from it economically. However,
there are some Cumbrians who ask visitors to their region
to look around and to see the harm that they believe BNFL
is doing, or may yet do. Among those who are not delighted
to have Sellafield as a neighbour in their midst are the
members of CORE (Cumbrians Opposed to a Radioactive
Environment).

CORE started life in 1980 as the Barrow Action Group, to
oppose the importation of spent nuclear fuel through the port
of Barrow-in-Furness for reprocessing at Sellafield. Since
then, CORE has widened its campaign to cover all aspects of
Sellafield's operations, including the radioactive sea and air
discharges, resultant contamination of the local environment,
and (what it claims is) the health detriment to local
communities and wildlife.

CORE has offered people its own 'Alternative Tour of
Sellafield', which highlights some of the local problems that
it says BNFL would rather ignore. It also maintains a website
(http://www.corecumbria.co.uk/). Campaign commitments
by CORE over the last decade have included the fight against
the operation of the new reprocessing plant (THORP) and the
successful battle against the proposal by the nuclear waste
agency Nirex to site a deep underground waste dump on the
edge of the Lake District National Park, for British and
foreign radioactive waste. The campaign against the

fabrication of MOX plutonium fuel and its transportation from Sellafield continues.

CORE describes itself as 'a non-political, non profit-making organisation, raising funds through membership fees, public donations, grants and consultancy fees. The group, with two full-time workers and several volunteer workers based in their Barrow-in-Furness office, works closely with other local, national and international environmental and anti-nuclear groups.'

Understanding Sellafield
Before exploring the history and development of Sellafield, and describing the new facilities of THORP and MOX in detail, it is worth recalling the initial sense of wonder and delight that surrounded the launch of nuclear power as an ostensibly clean and everlasting source of energy. It is also helpful to consider briefly the nature of some of the materials with which employees work at Sellafield, including uranium and plutonium, and to note that there is a broad range of customers for whose nuclear reactors they are reprocessing radioactive fuel. It is to these matters that we turn in the following chapter.

CHAPTER 3

Atomic Smitten

B eing on board was very exciting. *The Savannah* positively gleamed, its hard surfaces shining with the confident glow of a new age. To a boy such as myself in the early 1960s, the mere proximity of a nuclear reactor was thrilling. This wonderful US ship, visiting Dublin port, could run forever on atomic energy. Or so it seemed.

The twentieth century was smitten by the lure of the atom. The dropping of 'A-bombs' on Hiroshima and Nagasaki hastened the surrender of Japan at the end of World War II, and politicians convinced themselves that the mere threat of nuclear war might put an end to major international conflicts. Unsuccessful efforts were made to limit the spread of nuclear weapons beyond those great political powers that first developed them. At the same time, nuclear energy was perceived to be a boon for civilians and was advocated as a clean and safe source of virtually endless electrical power. On a trial basis, some non-military vessels were powered by their own small reactors. The first is thought to have been the icebreaker *Lenin*, built by the Soviet Union. *The Savannah* soon followed from the USA.

Not Impressed
There were those who saw from the start how discoveries about the atom were abused, how aggression and greed turned potentially valuable discoveries into weapons of mass

destruction and dirty energy. The 'Ban the Bomb' movement was among the first to protest, while international organisations such as Friends of the Earth and Greenpeace fight the nuclear industry today. There are big debates, and there are also some small but significant victories for those who oppose atomic energy. In the 1970s, a nuclear power station planned for Ireland was stopped by public outcry. Later, people in Northumberland, in the north-east of England, fought for years to prevent a new nuclear plant being built at Druridge Bay. Their success has been celebrated in a book by one of the leading Druridge campaigners (Gubbins, *Power at bay*).

Ships Scrapped

My boyhood excitement at being on board a vessel powered by nuclear fuel was not long shared by those who invested in the few civilian ships fitted with nuclear reactors. It is said that *The Savannah*, launched in the US in 1962, the *Otto Hahn*, launched by Germany in 1969, and the *Mutsu*, launched by Japan in 1967, all proved to be not economically viable. The construction of such ships was abandoned. Military vessels powered by nuclear reactors have been banned from entering many ports around the world. The love affair with the atom ended a long time ago. Today, the relationship resembles a marriage of convenience.

However, BNFL capitalised on the twentieth century's love affair with the atom. It continues to manufacture and recycle fuel for nuclear power stations located in the United Kingdom and overseas. The principal raw product that it uses to make that fuel is uranium, which has a suitable type of atomic structure. The atom is at the heart of nuclear energy.

Atoms

Every material thing in our universe is made up of tiny moving particles called atoms, which themselves vary in size. Uranium atoms are the largest natural atoms. Atoms contain even smaller moving particles that are called neutrons and protons. When an atom is split it produces energy and heat. This splitting is known as 'fission' and it occurs when a stray neutron strikes the centre of a uranium atom, its 'nucleus'. The scientific process of setting in motion a chain reaction, and of continuing fission within nuclear reactors, is central to the production of nuclear energy and is complex. However, the heat thus generated within a reactor is used in just the same way as heat from oil or coal might be, in order to transform water into steam, which then turns turbo-generators to produce electricity.

Uranium and Enriched Uranium
Uranium is, therefore, the basic necessity in the process of creating nuclear power. The ore is a heavy, silvery-white metal that is found in the earth of many countries. Having been mined, it is processed to become a fuel which is suitable for use in nuclear reactors. Uranium is as common as tin and far more common than gold. Most uranium mines are open-cast. The ore contains around 1.5 per cent uranium.

Natural uranium consists mainly of two types of atoms (isotopes): uranium-235 (U-235) and uranium-238 (U-238). Uranium-235 is the more important type because it undergoes fission more easily in a reactor, which is the process that creates heat and energy. The BNFL plant for manufacturing uranium fuel is at Springfields, Preston.

Natural uranium becomes more effective when it is 'enriched'. The scientific process of enriching uranium increases its uranium-235 content. Enriched uranium was not required for the older generation 'Magnox' reactors, but it is required for the more modern AGR (Advanced Gas-Cooled Reactors) and PWR (Pressurised Water Reactors) reactors. Enriched uranium is transformed into fuel pellets, which are then assembled in stainless steel cans to form a fuel pin. BNFL's plant for enriching uranium is at Capenhurst, near Chester.

In the early 1970s, the British, Dutch and German governments signed the Treaty of Almelo, an agreement under which the three partners would jointly develop the centrifuge process of uranium enrichment. This process involves sensitive technology and is co-ordinated by a body known as Urenco. In addition to the BNFL site at Capenhurst, it has enrichment facilities at Gronau in Germany and at Almelo in the Netherlands. Urenco claims that its nuclear sites 'have an environmental record second to none'.

The process of enrichment also leads to the production of some 'depleted uranium', which is uranium where the concentration of U-235 is considerably less than the amount found naturally. Depleted uranium is a high-density material that is sometimes used to manufacture a hard tip for conventional artillery weapons. The use of such shells during the Gulf War and the Balkans conflict has resulted in claims that civilians living in areas where they landed suffer from a far higher incidence of cancer than one would normally expect.

Plutonium and Pants

Plutonium is derived from uranium that has been used as nuclear fuel in a reactor. For decades, plutonium has been used to make nuclear weapons. But, today, it is also combined with uranium to make new mixed oxide (MOX) fuel for civilian nuclear reactors. In MOX reactors, plutonium substitutes for uranium-235 as the material that fissions and produces heat. Plutonium is enormously rich in energy. One gramme is said to contain more energy than two tonnes of coal.

In some remarkable passages of his official 1978 report on Windscale/Sellafield, Mr Justice Parker set out certain 'facts about plutonium' because he believed that there existed much misunderstanding about it. Among other things he stated (Parker, *The Windscale Inquiry*, p. 9) that, 'it is not true that plutonium is highly radioactive' and added that 'amounts could be eaten without appreciable harm'. In the most striking comment, he claimed that, 'As regards shielding from its radiation, it could be sat on safely by a person with no greater protection than, as Professor Fremlin put it, "a stout pair of jeans".'

Having made a case for the safety of plutonium, a case which at times seems somewhat petulant, Parker conceded that 'in certain circumstances plutonium is very dangerous to man' and that, if released into the environment, it 'persists for a very long time'.

For its part, BNFL today states that, 'Plutonium is radioactive, but the main type of radiation it gives off (alpha particles) is not very penetrating and does not pass through even thin layers of materials, such as thin rubber gloves.' However, BNFL acknowledges that, 'Plutonium is radioactive and highly toxic if you breathe it or eat it.' The company says that, 'We take very strict protection measures at each stage of the handling process to make sure the plutonium doesn't get into the body, by breathing or eating it, or through cuts in the skin.' BNFL notes that,

> There are materials in nature which give off more radiation than plutonium, such as the natural gas Radon, but they are generally less toxic than plutonium because of the way that the body deals with the material compared to plutonium. There are also many well-known substances used every day (like asbestos and hydrofluoric acid) which are very toxic and can be dangerous if they are not handled properly.

Recycling Plutonium

A recent official British estimate of the costs of storing plutonium into the future states that they are 'likely to be of the order of billions of pounds', and the majority of this liability will fall to the public sector' (Department for Environment, Food and Rural Affairs, *Managing radioactive waste safely*, p. 32).

So what can be done with the world's supply of plutonium, which is growing steadily as uranium is used as fuel in nuclear reactors? And what of the stockpile of plutonium from weapons made during the arms race between the USA and the former Soviet Union that have been, or are being, decommissioned?

In the United Kingdom plutonium is used in combination with processed uranium to make new fuels at Sellafield, the UK thereby becoming less dependent than otherwise on supplies of uranium and fossil fuels from abroad. As long ago as 1978, the Parker report officially declared the benefits for Britain of such reprocessing.

At present, BNFL does not recycle plutonium from the UK weapons programme. However, the company uses the existence of plutonium that has been recovered from spent fuel as a strong argument in favour of its controversial new Sellafield MOX plant. BNFL says that,

> Using MOX fuel helps us to manage the plutonium stockpile which would otherwise continue to grow if we only used conventional uranium fuel in reactors. Typical MOX fuel might contain between about 50 and 70 kilograms of plutonium for each tonne of fuel entering the reactor.

BNFL points out that

> Studies by the independent International Atomic Energy Agency have shown that if MOX fuel is burnt at reasonable levels, the world's stockpile of plutonium can be held steady and then gradually reduced during the next 10 years. But this is possible only if we maintain the skills and technology we already use in the reprocessing industry.

The alternative to recovering plutonium from used fuel and weapons and then recycling it to make new fuel is simply to store the plutonium. But, says BNFL, 'because the used fuel is around 96 per cent reusable uranium and around 1 per cent reusable plutonium, this option would waste an enormous source of future energy', as well as simply passing on the problem of disposing of it to future generations. So, argues BNFL, 'recycling plutonium through reprocessing plants is supporting and financing the development of MOX fuel technology today and the manufacturing processes that could change weapons into useful peacetime equipment tomorrow'.

Over 400 tonnes of MOX fuel, containing uranium and plutonium, are believed to have been safely loaded into reactors around the world since 1963. More than thirty reactors in Europe are licensed to use MOX fuel and it is intended for use in many more, especially in Japan. The production of MOX fuel at Sellafield is highly controversial and its transportation to reactors abroad by ship has already led to international protests.

Reactors

Nuclear reactors come in various shapes, sizes and generations. All sustain a continuous chain reaction within their nuclear fuel. By doing so, they create the energy and heat necessary to generate nuclear power. Once the heat is generated it is used, like oil or coal is in other plants, to heat water to steam. As noted above, this steam turns generators to produce electricity. Over the years, the design of reactors has improved significantly, bringing benefits in terms of safety and efficiency and enhancing the case of those who argue in favour of nuclear power.

Magnox (gas-cooled reactors) were an early model of reactor, invented in, and largely confined to, Britain. Their primary purpose was to produce plutonium for military purposes, but their added value as important suppliers of electricity in Britain gradually became more significant. The 'Magnox' tag derives from the fact that their fuel is natural uranium contained within magnesium alloy. A total of twenty-six Magnox reactors were built at eleven power station sites, seven in England and two each in Scotland and Wales. There are considerable differences in the detailed design of each of these power stations. The first four Magnox reactors were at Calder Hall, beside Sellafield in Cumbria. Some Magnox reactors are still generating electricity for the British grid. All of Britain's Magnox reactors are owned by BNFL. Magnox fuel is reprocessed at a special Magnox reprocessing facility at Sellafield.

AGR (Advanced Gas-Cooled Reactors) use slightly enriched uranium oxide pellets in stainless steel cans, cooled by carbon dioxide. Because of this design, they can reach a maximum temperature greater than that of Magnox reactors. AGRs were the second generation of British reactors, being unique to the United Kingdom, and were later transferred to British Energy, the privately owned power company. Spent AGR fuel is reprocessed at the THORP Plant at Sellafield, but the new MOX fuel cannot be used in AGR reactors. The prototype AGR was located at Windscale, Sellafield, where it is now in the process of being decommissioned.

PWR (Pressurised Water Reactors) are a common type of LWR (Light Water Reactor), being cooled by water, and are found around the world. Sizewell B, in Suffolk, is the only such reactor in Britain, and is operated by British Energy. It is the most modern type of nuclear reactor in the United Kingdom. Like AGRs, PWRs are also fuelled by enriched uranium, in this case uranium dioxide encased in a zirconium can. The PWR can run on MOX fuel.

BWR (Boiling Water Reactors) are less common than PWRs. BNFL has moved into manufacturing fuel for BWRs since acquiring the commercial nuclear reactor businesses of the multinational ABB (Asea Brown Boveri). The MOX plant at Sellafield produces fuel solely for PWRs and BWRs, thereby making a product that can be used in only one British nuclear power plant.

FBR (Fast Breeder Reactors) 'breed' their own replacement fuel. Indeed, when spent fuel comes out of an FBR it may contain enough plutonium to provide, after reprocessing, not only its own replacement but also some to spare. The surplus can be stored until there is enough for a whole new reactor. However, FBRs are controversial and plans to proceed with them in Britain have been scrapped. The Dounreay nuclear facility, in Caithness on the far north coast of Scotland, was established on a former naval base as the centre for UK fast reactor research. It is now being decommissioned.

The Advanced Passive 1000 (AP1000), designed by BNFL, is a 'state of the art' reactor which the company says 'makes nuclear power more affordable now than ever before'. The AP600, a smaller version of the AP1000, is already licensed in the US and the AP1000 may achieve US design certification by 2004. In September 2002, Norman Askew, BNFL's chief executive, said that, 'The AP1000 is ready for implementation now but to get electricity produced in 10 years time from the first of a new generation of nuclear reactors needs action today.' According to BNFL, generating 1,000 Mw (megawatt) of electricity from the AP1000 only requires 25 tonnes of nuclear fuel per year:

> To generate the equivalent amount of electricity in a coal fired power station would use around three million tonnes of coal. If the electricity provided by nuclear reactors worldwide were to be generated instead by gas-fired stations, CO_2 [carbon dioxide] emissions would increase by over one billion tonnes per year and this would double if coal stations were used. Replacing all the current UK nuclear capacity with AP1000 reactors would only add about 10 per cent to the UK's nuclear waste inventory over their lifetime.

The International Atomic Energy Agency (IAEA) notes in its latest annual report that, 'Many advanced reactor designs are in various stages of development in national research programmes around the world.'

The concept of prefabricated reactors is also being considered by a number of power companies at present. According to the International Atomic Energy Agency, several small to medium sized designs seek to benefit from modular structures and systems for rapid on-site installation and from economies of series production, as well as from their potential appeal for countries with small electricity grids or power needs in remote locations. They may also be appropriate for specific applications such as district heating, desalination and hydrogen production. Such uses would facilitate the expansion of nuclear power.

Bad Reaction
In his introduction to Michael Flood's report on the UK Atomic Energy Authority (UKAEA), published in 1988 by Friends of the Earth, Jonathan Porritt wrote scathingly that,

> If there is one single factor which accounts for the differences between the disgraceful record of the British nuclear industry, and the apparent success of the industry in France, it is surely the failure of British scientists to come up with and stay faithful to the right reactor design.

Porritt continued,

> The litany of failure in this respect must depress even the most enthusiastic advocate of nuclear power. Of the five systems the UKAEA has worked on, just two have been put into commercial production. One of those (Magnox) is now obsolete; the other (the Advanced Gas-Cooled Reactor) has been a financial disaster. And even after thirty years of intensive research and development, the Fast Breeder Reactor has got mighty little to show for itself.

It was the spectacular failure of one of Britain's earliest reactors that first brought Sellafield to the attention of the public in Britain and Ireland. That was when Sellafield was better known as 'Windscale'.

CHAPTER 4

Windscale

Windscale is the name of a bluff overlooking a small river, on the seaward side of Sellafield in Cumbria. 'Windscale' was also the name initially chosen by BNFL for its nuclear facility in this area, in order to distinguish clearly between the Sellafield site and the site of another BNFL nuclear facility at Springfields, near Preston in Lancashire. The plant at Springfields extracts uranium from ore and prepares it as nuclear fuel elements for Sellafield.

Nevertheless, the Windscale complex continued locally to be known also as 'Sellafield' and the latter name was eventually accepted for official use, instead of 'Windscale'. In the public mind, by then, the name 'Windscale' had come to be closely associated with the fire and leaks of radioactivity that occurred there in 1957.

Child of War
The nuclear facilities at Windscale/Sellafield were born out of the arms race. In 1945, as the United Kingdom emerged from the trauma of World War II, the British government devoted substantial resources to a nuclear research and development programme. It was believed that atomic weapons might not only deter aggression, but also put Britain in a much better position to secure the co-operation of the United States in military and strategic matters. The UK government also saw

possible benefits in the future application of research to the development of civil nuclear energy.

Following its experiences during the war, when Germany had landed missiles bearing conventional warheads on London, the UK government was both determined and desperate to develop nuclear weapons. In particular, the dropping of atomic bombs on Japan had brought home to the world the fierce reality of such armaments. Accordingly, the British gave an atomic weapons programme the highest priority. Locations were chosen as quickly as possible for the siting of various facilities needed to manufacture an atomic bomb. One of these sites was Sellafield, where a former munitions factory stood. In 1947, work began on developing this site.

Thus, Sellafield was first commissioned not as a civilian nuclear plant but as a nuclear weapons factory, where uranium fuel elements were to be irradiated in two atomic piles to produce plutonium and other atomic weapons materials. These earliest piles were never intended to serve any civilian purpose. They were not suitable for generating electricity. But they were fine for making plutonium for bombs. In 1949, with the plant at Windscale still incomplete, the government of the United Kingdom was deeply alarmed to learn that the Soviet Union had carried out its first known atomic bomb test. However, in October 1950, Pile No.1 at Windscale/Sellafield came into operation ('went critical'). In June 1951, Pile No.2 followed suit. It was to be at the first of these two piles, or 'stacks', that a fire broke out in 1957 and caused the worst known accident in the history of Britain's nuclear facilities.

In January 1952, the first irradiated fuel rods were processed at the chemical separation plant at Sellafield. Two months later, on 28 March 1952, the first piece of plutonium made in Britain emerged. It was despatched to the Pacific for the first British nuclear weapons test, code-named 'Hurricane'. This took place on 3 October 1952, when a device with a force equivalent to 20,000 tons of TNT was detonated in shallow waters far off the north-west coast of Australia.

Why Cumbria?
Sellafield was not the site that was first identified as being the most appropriate for producing plutonium for Britain's nuclear bomb programme, and for reprocessing and storing the wastes involved. Earlier, consideration had been given to Harlech, on the Welsh coast, and to a place between Arisaig and Morar on the west coast of Scotland.

If the United Kingdom had adopted criteria used earlier in the United States for determining the 'safest' distance between nuclear reactors and centres of population, then only a site on the north or west coast of Scotland would have been chosen. However, expediency and the adoption of new technology led to Sellafield being selected. Harlech was rejected as being too close to centres of population and as being of sensitive historical significance. Arisaig was rejected precisely because it was remote. To build near Arisaig would have required the management of a massive construction project, involving lengthy work on foundations, as well as the mobilisation of labour from distant centres of population. Britain was in a hurry to explode a nuclear device. So the government decided to look elsewhere. Furthermore, at this time, a new and ostensibly safer system of air-cooling reactors was designed, to replace water-cooling, and this fact was seized upon. While a very remote site was still considered desirable in case of catastrophe, the latest technological development was enough to persuade the authorities that the risk of picking somewhere closer to London and relatively less remote than the Highlands of Scotland was acceptable. The site chosen was the former munitions (Royal Ordnance) factory at Sellafield in Cumbria. The factory had been in production during World War II (1939–45).

Windscale 1947–1971: A Law unto Itself
In 1947, when it acquired the site of the old munitions factory at Sellafield, the Division of Atomic Energy of the United Kingdom Ministry of Supply wanted it, particularly, both for the construction of two air-cooled nuclear piles, or

reactors, for the production of plutonium for nuclear weapons, and for the building of a fuel reprocessing plant to facilitate the 'extraction' of plutonium. So, from the very start, reprocessing was a critical aspect of activities on the site. The reprocessing of nuclear fuels has taken place at Sellafield since 1952. The safety of reprocessing procedures is a cause of concern to this day.

Building began in earnest at Sellafield in September 1947. More than 5,000 men were employed to construct the nuclear complex, with high wages and steady overtime attracting labour to the coast. However, a number of serious design faults emerged in the process of construction. These were overcome to the satisfaction of the design team. In October 1950, as we have seen, the first of the two Windscale piles came into operation, or 'went critical'. In June 1951, the second did likewise. Then, in 1953, work commenced on building four more nuclear reactors at Calder Hall, adjacent to the new Windscale complex. In 1954, Windscale, including the Calder Hall reactors, was transferred from the Ministry of Supply to the UKAEA. The four carbon dioxide-cooled reactors at Calder Hall came into service between 1956 and 1959. Calder Hall was the world's first industrial-scale nuclear reactor. It was thought that cross-over benefits would ensue from developing both military and industrial nuclear programmes at the one location.

During this period, in 1957, fire broke out in one of the two, earlier, air-cooled reactors at Windscale. The fire resulted in an escape of radioactive particles which is thought to have had serious effects on the health of a small number of people in Britain and Ireland. The extent of those effects is still being contested today by the United Kingdom government and by BNFL. The reactors were disused after the fire, but have remained potentially dangerous.

In 1971, the Atomic Energy Act was passed. It established British Nuclear Fuels Ltd (BNFL), to which responsibility for the majority of the Windscale and Calder Hall sites was passed from the UKAEA that same year. Only then, in 1971, did Windscale and Calder Hall finally come to require licensing by the United Kingdom Nuclear Installations Inspectorate from which UKAEA sites had been and continued to be exempt.

For more than twenty years after its birth, the growing complex now known as Sellafield had been managed by Britain's atomic energy personnel largely as a world unto itself.

Piles of Danger

We have seen that the first two nuclear reactors built in Britain were those constructed at Windscale, alias Sellafield. These massive reactors, known then as 'piles' or 'stacks', were designed and built hastily. As we have also seen, speed was of the essence because the United Kingdom government wished to manufacture nuclear weapons as quickly as possible. The method employed to air-cool these first British piles was, in effect, experimental and would not be used subsequently in the nuclear energy programme. It involved blowing air straight through the core at atmospheric pressure. The four reactors that subsequently came into service between 1956 and 1959 at Calder Hall, adjacent to Windscale, were instead carbon dioxide-cooled ('gas-cooled').

Before constructing the first two air-cooled nuclear piles at Sellafield, scientists had considered adopting the new system of gas-cooling for those stacks, which they later adopted for Calder Hall, but their proposals were deferred in the interests of making haste. At that point, the weapons programme took priority. There were some major operational problems at the two air-cooled piles from the outset, including an extremely serious incident in May 1952, when 140 cartridges were displaced from the core and lodged in outlet ducts and elsewhere. Then, in 1957, fire broke out.

What Caused the Fire of 1957?

The fire of 1957 occurred in Pile No. 1, in the course of a controlled release of energy from its graphite blocks. This release was expected to be routine, involving a single heating of the pile. However, various difficulties had been encountered in conducting the procedure previously and, for that reason, it had been decided to carry out such releases less often than before.

On 7 October 1957, all necessary steps were taken to shut down the pile in anticipation of the planned release. Early on the morning of 8 October, something began to go wrong. Additional nuclear heat was then applied within the pile, because it was thought that the graphite temperatures necessary for a release of energy were not being achieved. On three earlier occasions, in the course of similar releases, more energy had been successfully applied to increase heating. However, the Penney Inquiry later declined to endorse claims by a Sellafield physicist and pile manager that, on this occasion, the general tendency of the graphite temperatures had been downwards. The Inquiry found that the primary cause of the accident was this additional nuclear heating: 'Having regard to the high temperatures already existing in parts of the pile at this time, the second nuclear heating was too soon and too rapidly applied', the report of the Inquiry reads.

During 9 October, the temperature within the pile became high enough to require cooling. By the morning of 10 October, staff suspected that there might be a problem. However, special scanning gear, utilised for the automated inspection of highly radioactive areas of the pile, was found to be jammed and could not be moved. It was not the first occasion on which the scanning equipment had become jammed. With the internal scanning gear out of action, it was decided to take samples of air coming from the stack and to test these for radioactive particles. The tests showed a positive large reading, which suggested that there had been a bad burst within the pile. The Penney Inquiry concluded that, most probably, the rapid increase in temperature of the fuel elements due to the second nuclear heating had caused the failure of one or more of the fuel element cans. The exposed uranium from inside the can, or cans, then oxidised and gave a further release of heat. This, together with the rising temperatures occasioned by later releases, initiated the fire.

Various efforts were made to bring the fire within the pile under control. The supply of steel push rods, used to discharge channels from the hot region, ran out. Early on the morning of 11 October, scaffolding poles were brought from a nearby construction site and used for pushing, instead of the usual steel rods! As the fire burned inside it, there arose a real danger that the whole pile might be ignited. Accordingly, management made a major decision to use water to extinguish the fire. The use of water carried with it inherent risks, including the possibility of a hydrogen-oxygen explosion. Employees at Sellafield who were not directly involved in the emergency were warned to stay indoors and to wear face masks. Water was then poured into the pile at a rate rising to 1,000 gallons per minute. The fire was thus successfully extinguished.

The Penney Inquiry found that a major technical defect contributing to the accident was an inadequacy of instrumentation for the safe and proper operation of a release of energy. Another serious defect was the absence of an appropriate operating manual. Moreover, in the condition of stagnant air which had necessarily been created within the pile, there was no means of detecting the smouldering uranium cartridges which were believed to have been a key event in the development of the accident. Penney also found that 'the operations staff at Windscale are not well supported in all respects by technical advice' and that 'the Windscale organisation is not strong enough to carry the heavy responsibilities at present laid upon it'.

The Penney Inquiry published its report promptly, on 26 October 1957. It observed that 'a worse accident was averted' and paid tribute to the way in which the Windscale staff had finally got on top of the crisis. While Penney found little evidence to suggest that there were any serious health consequences of the fire, either for Sellafield employees or for

the public at large, his report was by no means the last word
on the subject.

A late and costly addition to the Sellafield piles had been
the external filters known as 'Cockcroft's follies', named after
John Cockcroft of the United Kingdom Atomic Energy
Agency. These filters reduced harmful emissions during the
fire of 1957.

Effects of the Fire of 1957

To this day, controversy continues to erupt about the effects of
the fire at Windscale in 1957. Hewitt and Collier (*Introduction
to nuclear power*, 2000, p. 188), recently wrote that 'the
incident is of particular interest in nuclear safety analysis
because of the iodine release, which was much greater than
that which occurred, for example, at Three Mile Island'.

There is no doubt that the accident and its consequences
might have been much worse than they actually were. This
incident was far less damaging than the later Chernobyl
disaster. Just how close it came to being much worse is a
matter for conjecture, but the fire at Windscale in 1957
served even then as a wake-up call for those who were
complacent about the dangers of nuclear power. Measured
dosages of radiation received by individuals in the Sellafield
complex and surrounding areas, during and following the fire
of 1957, appeared at first sight to have been relatively
modest. Local milk supplies were destroyed for a period in
order to prevent their consumption, particularly by children.
Nevertheless, there were to be later reports of higher than
average rates of leukaemia in children in Cumbria and along
the east coast of Ireland, and it has been alleged that any such
higher rates are related to the fire at Windscale in 1957.

In 1957, immediately after the fire at Windscale, special
attention was paid to the effects of the escape of radioactive
iodine vapour from Pile No. 1. The deposition of iodine-131
on grass that is eaten by cattle is quickly carried to their milk.
A major part of the effort of the teams assessing activity in
and around Sellafield in October 1957 was devoted to milk

analysis for iodine. Children have small thyroid glands and often drink relatively large quantities of milk. They are especially vulnerable to iodine-131.

For a period, immediately after the 1957 fire, certain farmers were prevented from making milk deliveries. These farmers worked along a coastal strip of Cumbria approximately thirty miles long, ten miles broad at the southern end and six miles broad in the north. Their cows' milk was then poured down drains or into ditches. Samples of milk were taken also around the Lancashire coast, the north Wales coast, in the Isle of Man, and into Yorkshire and the south of Scotland. On the basis of these samples, it was thought unnecessary to extend the boundaries of restriction on milk supplies. Other possible sources of ingestion hazard were also examined in the locality, in particular vegetables, eggs, meat and water supplies. None of those were found to be harmful at the time. Among other fission products released had been caesium-137, strontium-90 and tellurium-132.

The Penney Inquiry heard evidence of high levels of radioactivity on grass and clothing in nearby Seascale, and on the clothes of people cycling to work along the track from Seascale on the morning of 11 October 1957. However, noted Penney, 'the activity measured on the clothing of the two witnesses mentioned above was some twenty times lower than that which would have constituted any hazard in accordance with the standards observed by the [United Kingdom Atomic Energy] Authority and based on the Medical Research Council tolerance levels'.

At first sight then, the fire of 1957 turned out to be just a warning of what might be. The results of tests were all very reassuring, even for those employees of Sellafield who were closest to the fire itself. However, there remain unproven suspicions that the effects of the fire were worse than was officially recognised. It later emerged that a radioisotope known as polonium-210 had been released by the Windscale fire. A suggestion that there was an official cover-up of this ingredient among Britain's secret atomic weapons material has been dismissed as 'implausible' by Lorna Arnold. However, she adds that 'it is incomprehensible that it should have been omitted' from a number of early reports on the fire (Arnold, *Windscale 1957*, p. 140).

In 1983, a British television documentary on Windscale raised the possibility of a leukaemia 'cluster' in Seascale, near Sellafield. But experts argued that, between 1950 and 1980, radiation doses from Windscale were well below what the population received from natural sources and from the fall-out from nuclear tests. Any such local cluster may be connected to the fact that one of the parents of each such child worked at Sellafield.

People living on the east coast of Ireland also claim to have suffered ill effects from the Windscale fire of 1957. They were assured from Britain that the low levels of dose released by the fire were highly unlikely to have had such distant effects. Allegations and counter-allegations about the possible effects of the fire of 1957 continue to be made. A suggestion that it resulted, in County Louth, in more babies than usual being born with Down's syndrome has been strongly rebutted in a study by Dean and others (See p. 262 below).

Winds of 1957

The Penney Inquiry reported on the direction of winds at Sellafield on 10 and 11 October 1957, when the fire in Pile No. 1 led to an escape of radioactive vapour. Such vapours form a plume which, in the case of Sellafield, might cross Glasgow or Edinburgh, Leeds or Manchester, or Ireland, depending on the wind.

Penney states that, 'Throughout Thursday 10 October, the ground wind was light but mainly off-shore, i.e. N.E. or N.N.E. During the night it changed to N.N.W. and throughout Friday a ten-knot wind blew, mainly N.W. and N.N.W., i.e. down the coast. Still later it appears to have changed to a S.W. direction.' This suggests that it blew lightly towards the north of Wales and the Irish coast on Thursday, before switching to blow stronger towards the English Midlands and then, changing again later, towards Scotland. However, winds at a certain height are said to have possibly been blowing in a different direction from the direction of the ground winds. This does not paint a very clear picture of the possible fall-out path. Nevertheless, apart from certain measurements taken at Belfast, there appears to be no evidence that the fall-out from Windscale reached Ireland by air, while there is evidence to show that it drifted over Britain. That week, the winds favoured Ireland.

Immediately after the fire of 1957, as much radioactive fuel as possible was removed carefully from the two air-cooled reactors. However, because of the nature of radioactivity, it was not possible simply to dismantle the structures, and the piles remained standing and dangerous. It was assumed that eventual 'decommissioning' would become easier in time, with the gradual decay of any radioactivity of short and medium half-life. By 1987, it became clear that the two reactors remained dangerous, notwithstanding the fact that they were no longer being used for any productive purpose. A structural survey disclosed defects in the chimneys and certain remedial work was then undertaken.

In 1994, an official United Kingdom Advisory Committee on the Safety of Nuclear Installations (ACSNI) found that the two old Windscale piles fell 'far below the "as safe as reasonably achievable [ALARA]" criterion' that is required of the industry (this criterion will be discussed in Chapter 16). Further work on the piles was undertaken, and their chimneys (or 'stacks') have now been largely taken down. The reactors themselves are being investigated and it is hoped to dismantle and decommission them entirely.

Lorna Arnold's Windscale 1957
During 1992, the first edition of a special study of the fire at Windscale in 1957 was published. Written by Lorna Arnold, the book tells the story of the men who designed, built and operated the reactors at Sellafield. In it, the author describes

the fire as 'the world's first major reactor incident' and considers its causes, consequences, effects and political importance. In 1995, a second edition of her book appeared.

One might expect the book to be deeply partisan for, at the time of its appearance, Arnold was a consultant to the UKAEA. Although the book was published by Macmillan Press, the copyright in its text belongs to the UKAEA, for which Lorna Arnold long worked on staff. She joined the health and safety branch of the UKAEA fifteen months after the fire of 1957, becoming its record officer in 1967. She had assisted the historian Margaret Gowing in writing that author's *Independence and deterrence: Britain and atomic energy 1945–52*, and, in 1987, herself wrote *A very special relationship: British atomic weapon trials in Australia*. In fact, her book on Windscale is a very useful and interesting starting point for anyone who wishes to learn more about the fire of 1957. Arnold has managed to throw new light on the episode and raises pertinent questions about secrecy and responsibility. Arnold devotes much attention to all aspects of the Windscale fire and its consequences up until the time that she was writing. Her book also contains useful appendices, including a full transcript of the findings of the Penney Inquiry.

Reviewing *Windscale* for *The Guardian*, Walt Patterson described it as 'exemplary'. He wrote that, 'Mrs Arnold has the rare gift of writing incisive, gripping narrative . . . Individuals leap from the page . . . Scrupulous and fascinating detail.'

Windscale: 1957–81
Before 1971, at Windscale, the only incidents of error that its managers were obliged to report to the government were any which involved a loss of life, or serious injury or 'significant' radiological effects beyond the boundary of the site. Between 1950 and 1957, there had been some escape of radioactivity from Windscale, although the true incidence and extent of any such releases may never be known. When tests were carried out after the fire at Windscale in 1957, for example,

it was found that the highest contamination by strontium-90 was in a few fields near the piles. This contamination had been mainly caused by previously undisclosed emissions of fairly coarse particles of nuclear fuel from the Windscale pile stacks in the years before the accident. Clearly, the filters fitted to the piles were not completely adequate.

By 1971, notwithstanding the scare of 1957, the standard of the plants at Windscale 'had deteriorated to an unsatisfactory level'. So found an official report by the United Kingdom Health and Safety Executive, published in 1981. Between 1950 and late 1976, there was a total of 177 listed 'incidents' at Windscale, these being caused by procedural errors, personal errors and design errors. The incidents involved either personal exposure to radiation or contamination in excess of statutory limits, or an abnormal release of radioactivity into the environment, or a breach of criticality clearance procedures. (Paper submitted by BNFL to the Windscale Inquiry, see Charlesworth et al., *Windscale: the management of safety*, 1981 p. 11).

In February 1977, radioactive liquid leaked into the ground from a silo for the storage of irradiated Magnox at Windscale. Later that month, being increasingly concerned about safety at Windscale and other facilities, the United Kingdom government commenced a new system of reporting on incidents occurring at all licensed nuclear installations. The reports, published four times each year by the official Health and Safety Executive, reveal that half of all incidents reported during the following two years occurred at Windscale. Two incidents involved dangerous releases of radioactive materials into the ground. Neither involved the public, it is said.

In 1979, the United Kingdom's Health and Safety Executive undertook a major review of safety and of the plant at Windscale, and it was this report that was published in 1981. The inspectors found that BNFL 'did not always address the associated managerial or procedural deficiencies which might have been contributory factors' to particular incidents. The report states that much of the modification and refurbishment work then being carried out by BNFL at Windscale 'represented a catching-up with the past'. It noted that

> this work was necessary because of insufficient investment in plant and equipment in previous years, one consequence of which was a growing shortage of adequate facilities for reprocessing and associated work. (Cited in Charlesworth *et al.*, *Windscale: the management of safety*, 1981 pp. 5, 12)

The authors of the 1981 report on Windscale made many recommendations for improvements in safety but added that,

> we would not like this to give the impression that we regard Windscale as a dangerous place at which to work or near which to live. In our view the Company (BNFL) shares with the best of the British chemical industry the merit of devoting considerable attention to health and safety.

However, notwithstanding this endorsement of BNFL, the report went on to give a worrying account of the past management of the Windscale/Sellafield site. It said that,

> The general picture presented by the reports [safety assessment reports 1974–7] is one of past neglect, including the normal maintenance of buildings, and of occasional undermanning; there had been a general lack of expenditure on repair and maintenance. Many of the recommendations in the reports were directed to quite elementary matters which as a matter of good practice should have been put right. There is ample evidence to show that the general standards of housekeeping on the site were low, and although a great deal has since been accomplished, there are still pockets of the site needing attention [in 1981].

Calder Hall and Chapelcross

In October 1956, Queen Elizabeth II ceremonially opened the first of four nuclear reactors at Calder Hall power station. Calder Hall is not only regarded as the original commercial nuclear power plant but it became the longest operating one in Britain. It has provided plutonium for the military, as well as generating electricity for the Sellafield site and, to a limited extent, for the national grid. Because of the early nature of its design, the potential risks to public health posed by a serious accident at Calder Hall have been considerably greater than the potential risks from a similar accident at more modern plants.

Four reactors like those at Calder Hall were also constructed just across the Solway Firth. They stand at Chapelcross in Scotland, between Dumfries and Carlisle. Chapelcross, like Calder Hall, was intended primarily to produce plutonium as a by-product for weapons use. However, the military function of both plants became much less important over time.

In 1955, the official adoption of a ten-year 'Programme for Nuclear Power' involved the electricity generating boards in building a series of reactors to produce up to 2,000 megawatts of nuclear power by 1965. The programme was later expanded. These other early British nuclear power stations were at Berkeley, Bradwell, Hunterston 'A', Trawsfynydd, Hinkley Point 'A', Dungeness 'A', Sizewell 'A', Oldbury and Wylfa. The British also exported reactors to Italy and Japan, thus demonstrating that export revenue could be earned by the development of the industry in Britain.

Until 1971, Calder Hall and Chapelcross stations continued to be the responsibility of the UKAEA and were exempt from licensing. In 1971, they became the responsibility of BNFL and subject to licensing.

Calder Hall's four reactors had been expected to remain operational for less than thirty years but, later, were found to be serviceable beyond that. In 1996, the reactors were discovered to be generating low level liquid waste consisting

of water that contained the radioactive nuclide (or
'radionuclide') tritium. This was treated at Sellafield before
being disposed of in the Irish Sea. In 2001, there was a major
incident at Chapelcross when irradiated Magnox fuel
elements were dropped down a discharge well.

Until recently, it was intended that Calder Hall would cease
generating electricity between 2006 and 2008, Chapelcross
between 2008 and 2010. However, for financial and security
reasons, all eight reactors are now being shut down sooner
than expected. Three of the four reactors at Calder Hill are
already out of service and the fourth is expected to close
before the middle of 2003. The process of 'decommissioning'
these radioactive buildings can then commence.

To provide the power and steam requirements of Sellafield,
which the reactors of Calder Hall have met until now, a new
electricity generating plant fuelled by gas is being
constructed. This is being built off-site because, BNFL says,
the usual discharges from such gas stations do not conform
to the standards of discharge authorised for the Sellafield site
itself.

CHAPTER 5

Sellafield and BNFL

Since 1971, BNFL has run most of the facilities at Windscale/Sellafield. The site itself is a sprawling, industrial complex, which extends about two kilometres in length and two and a half kilometres in width. It is a maze of grey and brown buildings, some of which are more than fifty years old. Many look like factory units or large sheds. Today, its most outstanding feature is the massive bulk of the THORP reprocessing plant.

During the 1840s, Mannix and Whellan wrote that '*Sella Park*, or *Field*, is a hamlet, near the sea, five miles south of Egremont. It was formerly the property of the monks of Calder Abbey [nearby], who had here a deer-park.' Today, however, Sellafield is far from being a deer-park. Set in a patchwork of green fields are 600 acres of danger waiting to become a disaster. Most of its workforce are highly conscientious and are dedicated to preserving high standards of care and maintenance. Nevertheless, over the years, a number of incidents have occurred which demonstrate that it is not possible to expect perfection from any workforce, especially one as large as that at Sellafield where approximately 10,500 people are employed.

If accidents are inevitable, their consequences can be catastrophic. Secrets from Sellafield's military and industrial past may still lie hidden among its many structures. It is certain that the complex retains a substantial volume of highly radioactive waste that awaits storage in some future permanent repository, at a site not yet chosen by the government of the United Kingdom. Some other wastes are released down pipes that sneak through a gully at the back of Sellafield and pollute the marine environment.

Drinks Anybody?
Sellafield's 'liquor' is not recommended as part of a good night out. This particular Cumbrian liquor is brewed during the reprocessing of spent fuel from nuclear power plants. The stream of waste that is produced must be stored indefinitely. Drink it and you are dead. To put it more formally, Radioactive Liquid High Level Waste, otherwise known as Highly Active Liquor (HAL), is produced by the reprocessing of irradiated nuclear fuel at Sellafield. The waste, or 'liquor', has been accumulating from the 1950s and is kept at Sellafield, much of it in a facility known as B215. The liquor is a concentrated solution of fission products in nitric acid and is retained in a number of water-cooled 'Highly Active Storage Tanks'. There are 1,500 cubic metres of this particular liquor at Sellafield, which amounts to around 300,000 gallons, or the same volume as that of the water in a fifty-metre swimming pool.

According to one recent report for the European Parliament, published in 2001,

> The risk potential of certain hazards at Sellafield is very large. Liquid high level wastes currently stored at Sellafield contain about 7 million TBq (2,100 kg) of caesium-137, which is about 80 times the amount released through the 1986 Chernobyl accident. Assuming a 50 per cent release of caesium-137 in an accident at Sellafield, population dose commitment would range up to tens of millions of person-Sv (sievert) resulting in over a million fatal cancer cases.

This controversial report is known as the 'WISE' report, because the consultants responsible for its completion were associated with the World Information Service on Energy ('WISE') in Paris.

British officials have set an upper limit of 1,575 cubic metres for the total amount of high-level radioactive liquid that may be stored at Sellafield. Both the Health and Safety Executive of the United Kingdom and BNFL itself say that they wish to have all of this very dangerous waste converted into a solid and much safer form 'as soon as is reasonably practical'. The solid form is known as 'glass' (borosilicate glass) and is cast in stainless steel containers for keeping in the Vitrified Product Store. Storage of nuclear waste as a glass has advantages over its storage as liquor because the fission products in it are immobilised in the solid matrix, and the glass is cooled by the natural circulation of air. Keeping liquid cool, on the other hand, will continue to depend on the availability of installed services such as electricity and water. Those services can break down. The process of vitrification is proceeding far slower than anticipated, because of technical difficulties, and much liquor remains to be vitrified. It is hoped to reach a storage level of 200 cubic metres by 2015.

If nuclear liquor is not kept cool, it can boil. Boiling liquor can overload the ventilation filtration system and lead to a nuclear accident. A number of important engineered containment barriers keep the radioactive liquor in its tanks at Sellafield. The United Kingdom's Health and Safety Executive recognises that if these barriers were not maintained there could be 'radiological consequences both on and off the site'. The consequences could be extremely serious.

A Bad February
It was a case of three strikes but you are *not* out! In February 2000, the Nuclear Installations Inspectorate (NII) of the HSE published three reports on aspects of Sellafield and their findings gave serious cause for concern. The reports certainly did nothing to convince critics that the nuclear industry is one that can be entirely trusted to keep its own house in order.

One report raised eyebrows, in particular, because it involved dishonesty in the keeping of records. It was entitled *An investigation into the falsification of pellet diameter data in the MOX Demonstration Facility at the BNFL Sellafield site and the effect of this on the safety of MOX fuel in use*. This report, concerning false data, made people wonder what else might be happening at Sellafield about which they had been deceived. The second report of February 2000 concerned *The storage of liquid high level waste at BNFL*. Here again, the NII found room for improvement. This report was later updated, in August 2001. The third report of February 2000 comprised a *Health and Safety Executive team inspection of the control and supervision of operations at BNFL's Sellafield site*. Weaknesses were found 'which showed that there had been a deterioration in safety performance at Sellafield'. All three reports, and the update, may be downloaded from the site of the Nuclear Safety Directorate of the Health and Safety Executive (http://www.hse.gov.uk/nsd). The reports alarmed the Irish government, amongst others, and were mentioned by a government minister in both houses of the Oireachtas (Irish parliament).

Since the reports were published, BNFL has been eager to demonstrate its willingness to mend its ways and to implement the recommendations of the NII in each case. In BNFL's *Annual Report 2001*, Chief Executive Norman Askew stated that,

> The HSE has judged that we have successfully implemented all recommendations in two of the three reports. Doing so at the MOX Demonstration Facility in such a concentrated time span was a particularly noteworthy achievement. We also addressed all of the recommendations in the report on the tanks storing Highly Active Liquor (HAL) – a liquid waste recovered from spent fuel reprocessing – at Sellafield. The HSE fully endorsed the actions we took and agreed a programme to reduce stock levels in the future.

Askew added that,

> Regarding the lengthier Team Inspection Report, which addressed issues across the whole Sellafield site, work has progressed well. The HSE has formally endorsed more than half of the 41 key programmes of work linked to the recommendations. A progress report published by the HSE in February 2001 stated 'we have found evidence to support the claims that both the management and workforce are determined to ensure that the improved systems are effectively implemented'.

Wise Words

The authors of the controversial 'WISE' report for the European Parliament were concerned that, although gaseous releases of most nuclides from Sellafield have not varied to a marked extent since the 1970s, 'iodine 129 emissions have increased 10-fold'. They acknowledged that radioactive marine releases of carbon-14, strontium-90 and caesium had declined markedly in the early 1980s, but found that in the mid-1990s increases occurred in releases of carbon-14, cobalt-60, strontium-90, technetium-99 and iodine-129. They claimed that internal BNFL documents suggest significant increases in nuclide releases in the future at Sellafield and they reported that, 'For some *"worst case"* scenarios, the operator predicts *"levels approaching or above the limits"* for sea discharges of over half the currently authorised radionuclides. A similar situation is expected for aerial releases.'

The 'WISE' consultants concluded that the increase in releases of key radionuclides from Sellafield in the late 1990s, and expected future discharges, were and will be inconsistent with the United Kingdom's obligations under the OSPAR Convention. They stated that the deposition of plutonium within twenty kilometres of Sellafield attributable to aerial emissions had been estimated at 160–280 GBq (billion becquerels), 'that is two or three times the plutonium fallout from all atmospheric nuclear weapons testing'. They estimated that between 250 and 500 kilograms of plutonium from Sellafield has been absorbed as sediments on the bed of the Irish Sea and claimed that this plutonium 'represents a long-term hazard of largely unknown proportions'.

The 'WISE' report has been criticised in Britain and has been treated with great caution by the European Commission, following observations made on it by independent scientists, who included Professor P.I. Mitchell of Dublin. In March 2002, the National Radiological Protection Board of the United Kingdom claimed that 'some of the conclusions drawn in the report are based on erroneous comparisons or reflect a lack of objectivity'.

What Exactly Is BNFL?
BNFL was created in 1971 to operate the fuel enrichment, fuel manufacturing and fuel reprocessing services developed by the United Kingdom Atomic Energy Authority (UKAEA) and to run, also, the Calder Hall and Chapelcross reactors. It was set up as a limited liability commercial company with all of the shares held by the UKAEA on behalf of the United Kingdom's Secretary of State for Energy. BNFL points out that much of its plant and equipment was already in existence in 1971, and that earlier activities burdened the company with a legacy of problems. BNFL's annual report for 2001 notes that,

> Those activities stemmed from a military programme during the cold war, and the priority when the plants were built was to get them operational and effective as quickly as possible. When these plants were designed, limited thought was given to their ultimate decommissioning and demolition and as a result the costs of so doing now are very considerable.

BNFL is still wholly owned by the state and is still responsible for all of Britain's oldest nuclear power stations. These 'Magnox' reactors are progressively being closed down. However, Britain's somewhat more modern nuclear power plants were privatised in the 1990s and became the property of British Energy plc (BE). BE has been a major customer of BNFL, for the supply of fuel, reprocessing services and the storage of spent fuel.

While BNFL is owned by the state, the government intends for it eventually to seek private sector investment. Still known as 'BNFL', since the days when the simple tag 'Ltd' ('Limited') was affixed rather than 'plc' (private limited company), it is a big and ambitious multinational organisation. BNFL has its corporate headquarters at Risley in Cheshire. As already noted, it makes nuclear fuels at Springfields, Lancashire, and at Sellafield, Cumbria. Sellafield is also its principal storage and reprocessing plant. It is intended by the government that BNFL will shed its ownership of the Sellafield site to a new Liabilities Management Authority (LMA), which will take on the financial responsibility for various public sector civil nuclear liabilities and assets. In the meantime, Sellafield itself remains firmly the property and the responsibility of BNFL.

BNFL employs 23,000 people in sixteen countries, with more than a quarter of its employees working at Sellafield alone. It now owns Westinghouse, the American nuclear energy producer, and boasts of currently having 12 per cent of the world's nuclear market. It also recently acquired the nuclear businesses of the multinational ABB (Asea Brown Boveri), and provides services to other companies in the nuclear power industry. BNFL describes itself as one of two key players in the global spent-fuel management and recycling services market, with Sellafield being central to those services. It produces fuel for all of Britain's nuclear power stations. It is also a leading supplier of fuel for generating nuclear power in other countries. The company claims to be one of only two companies in the world able to provide a complete nuclear fuel cycle service, from the manufacture and enrichment of uranium fuel through transport and recycling to the management of waste.

BNFL runs a substantial programme of investment in research. For example, the company has strong links with scientific centres at the universities of Manchester, Leeds and Sheffield, and is participating in a dozen European Union research projects. It also has a community relations programme. For example, during 2000–2001, BNFL contributed nearly £3 million sterling to economic projects in Cumbria. Employee involvement in community activities, across its multinational operations, has included a 'Day of Caring' project in Pittsburg, Pennsylvania.

For operational purposes, until recently, BNFL divided itself into four principal divisions. These were

Westinghouse (for nuclear fuel manufacture and reactor services)

Magnox Generation (managing Britain's older nuclear power stations and one hydro-electric power station for England and Wales)

Spent Fuel and Engineering (managing Sellafield and BNFL's engineering resources and, at the start of 2002, employing more than 7,600 people)

Nuclear Decommissioning and Clean-Up (managing the dismantling of out-of-date nuclear facilities)

BNFL owns a number of subsidiaries. These include Direct Rail Services (DRS) and Pacific Nuclear Transport Ltd (PNTL), both of which transport radioactive materials belonging to BNFL and its customers to and from Sellafield. Details of the many British and overseas nuclear sites owned by BNFL companies, including more than two dozen in the United States alone, are available with much other information on the BNFL website at www.bnfl.com.

Homer Who?
Springfields, Lancashire, has been the home of nuclear fuel manufacturing in the United Kingdom since 1946. It is at Springfields that BNFL turns uranium into the stuff of nuclear power. Under no circumstances ought Springfields, England, to be confused with that Springfield, USA, which is the fictional home of Homer Simpson and his family. Remarkably, there are those who say that they have never seen *The Simpsons*

on television, and they include at least one senior BNFL manager. They may not know that the fat and lovable Homer is depicted as an employee of the local nuclear power plant, whose work practices leave something to be desired when it comes to care and safety. Any resemblance between Homer's employer and BNFL is entirely coincidental, and creator Matt Groening could not possibly have had in mind the United Kingdom's nuclear power programme when he located his Simpson family in a town called Springfield. Could he? There are many towns called Springfield in the United States, although the Simpsons' home is never identified as being in any particular state. Besides, the English town has an 's' at the end of its name. D'oh . . .

Restructuring

In July 2002, BNFL announced that it was reorganising its business into just two subdivisions, instead of four. This decision had been taken in anticipation of the process whereby a Liabilities Management Authority will be created by the state to take over much of the financial and legal responsibility for the Sellafield site. It was also in anticipation of BNFL itself seeking investment in order to recreate itself as a Public Private Partnership. However, although BNFL was streamlined internally from four to two groups, operations at Sellafield were actually divided between these two entities. Thus, THORP, Magnox reprocessing and the MOX fuel business, along with BNFL's marine transport activities, are now being provided by the 'Nuclear Utilities Business Group'. Management and operations activities at Sellafield (including historic waste management) are being provided by the 'Government Services Business Group'.

An Insider's View

In 1996, *Inside Sellafield* was published by Quartet Books. Written by Harold Bolter, formerly the longest serving director of Sellafield, the book raises questions about the way the nuclear industry is managed. Bolter had long had responsibility for public information, security, political liaison and community relations at Sellafield. Despite having spent half a lifetime at the centre of the controversies which the site attracted, Bolter came to lose some of his early enthusiasm for nuclear power.

Through the pages of *Inside Sellafield*, Bolter chose to speak out about what he believed to be a record of inefficiency, government interference, lack of vision and, above all, a culture of concealment. He thought this culture had contributed greatly to a perception of the nuclear industry as the most feared and mistrusted of all sources of power. Bolter had broken with BNFL in difficult circumstances in 1993, after being cleared of financial misconduct in connection with payment for work done at his home by a contractor who also worked for BNFL. He claimed to have contemplated suicide because of the way he felt he had been treated. But, instead, he wrote about his experiences within the company and about the self-imposed secrecy among managers within Sellafield. His book is useful for anyone trying to understand how and why Sellafield operates. Bolter writes that its continued functioning is crucial to Britain's nuclear policy. 'Shut down Sellafield', he says, 'and the nuclear industry would quickly grind to a halt'.

In 1996, Bolter also featured in a television documentary made by Border TV for Britain's ITV network. Also called 'Inside Sellafield', the documentary examined twenty-five years of nuclear reprocessing in the UK and asked what future there is for the industry. A transcript of the programme may be viewed on the web at http://www.border-tv.com/inside/inside.html.

A New Kind of BNFL?

When the government of the United Kingdom proceeds to establish a Liabilities Management Authority, as it intends to do, there will be very significant implications for BNFL. The LMA will relieve BNFL of most of its old properties in Britain, including Sellafield, and take over the massive liability of caring for the contaminated buildings and stored waste of the past. Then, a 'New BNFL' will be freed to compete for the contracts to service such sites. It will also be free to attract private investment into the company in a way in which it is not at present able to do because of the burdens that it carries from the past. Thus, under such a restructuring, 'New BNFL' may pursue its growing international business commitments while continuing to manage British sites. However, it would manage them under contract to the Liabilities Management Authority and would no longer be obliged to shoulder responsibility for a dirty legacy of buildings and old waste. BNFL points out that much of what exists on the Sellafield site, for example, was created under early civilian and military research programmes and has been there since before BNFL itself came into existence.

BNFL's chief executive, Norman Askew, immediately welcomed the Department of Trade and Industry's White Paper proposing a Liabilities Management Authority. He said that, 'we all look forward to embracing the challenge of a world where we operate in a more commercially competitive environment . . .' It remains to be seen to what extent, exactly, private investors will be willing to invest in BNFL once it sheds its burden of care for much of Sellafield and for the old Magnox stations. There may well be considerable interest in the company, particularly in light of its overseas client base for reprocessing.

Doesn't Matter
Restructured or not restructured, BNFL does not impress the Irish government when it comes to its continued involvement in the operation of Sellafield. On 10 October 2001, Joe Jacob, its junior minister responsible for nuclear matters, said that,

> This Government is totally opposed to the continued operation and any expansion of the Sellafield plant.

He added that,

> The Government's concerns about the existing operations at Sellafield relate mainly to –

- the management of safety at the site
- the storage in liquid form of high level radioactive waste
- the continued reprocessing of spent nuclear fuel at the site
- the continued operation of the old magnox reactors
- the discharge of radioactive waste to the sea
- the risk of a catastrophic accident.

Jacob said that,

A dominant theme of the Irish Government's campaign against Sellafield has been the genuine fear of an accident at the site. The multiplicity of existing operations at Sellafield increases the risk of a major accident. This risk has now become much more acute in the light of the horrific events of September 11th. Even countries with nuclear programmes are questioning the merits of reprocessing on economic and safety grounds. The nuclear spent fuel reprocessing industry is now separating more plutonium than the nuclear industry is able to absorb. Sellafield now has a stockpile of plutonium, which presents a potential risk to existing and future generations. The existence of such a large stockpile demonstrates how ill-advised it is to persist with reprocessing.

CHAPTER 6

Dumping in the Sea

It may be history. However, unfortunately, historical contamination is still affecting us. What was dumped from Sellafield into the Irish Sea, even decades ago, has left a radioactive residue that is still today contaminating the marine environment. The creation of BNFL, and the fact of its taking responsibility for Sellafield, did not stop the dumping. In fact, BNFL was specifically authorised to continue discharging liquid radioactive waste from Sellafield, through pipelines, into the Irish Sea. The authorisations from the Ministry of Agriculture, Fisheries and Food and the Department of the Environment did require BNFL to sample and to keep records of the waste so discharged. One hopes that they kept such records without failure.

One particular incident at Sellafield suggests that breaches of licensing and authorisation conditions may have gone unnoticed during the 1980s. In November 1983, a rubber dinghy being used during a Greenpeace protest became contaminated. It emerged that certain radioactive liquids were being flushed into the Irish Sea in an unacceptable manner. The NII found that relevant safeguards had failed. This was due, apparently, to over-reliance on particular administrative arrangements. One wonders if other irregular releases have gone unnoticed down the years, in the absence of other Greenpeace dinghies.

Recently, as was seen above, the authors of the 'WISE'

report for the European Parliament expressed concern about continuing and future authorised discharges from Sellafield. However, while Sellafield is the primary source of radioactive pollution of the Irish Sea, the levels of radioactivity do not appear to be such that they pose a significant danger to people swimming in or living beside it.

Improvements
Over the years, BNFL has taken steps to limit its radioactive releases from Sellafield and the total volume of such releases is said by that company to have decreased, comparatively. BNFL states that it has invested 'hundreds of millions of pounds to treat this "slightly contaminated water" and to reduce liquid discharges'. According to Cassidy and Green's 1993 book, *Sellafield: the contaminated legacy*, early discharges at Sellafield formed part of a deliberate experiment to examine the fate of radionuclides in the environment. The behaviour of radionuclides was then poorly understood. The authors claim that Sellafield did not at first adopt a precautionary approach but, instead, its operators incorrectly assumed that the discharged radioactivity would be safely diluted and dispersed in the Irish Sea. That assumption was wrong. Measurements today still clearly show that the 'dilute and disperse' policy resulted in widespread contamination.

In 1984, the United Kingdom's Ministry of Agriculture, Fisheries and Food reported that, since 1972, there had been 'a progressive reduction in the level of radioactivity in the liquid discharges from Sellafield' into the wider environment. Such reductions, said the ministry, were 'being achieved by the progressive introduction of new plant and the continued scrutiny of management procedures to ensure that the best practicable means (BPM) of limiting discharges are employed'.

By 1984, a Site Ion Exchange Effluent Plant (SIXEP), and a salt evaporator, had been brought into operation at Sellafield: 'Between them they are enabling substantial reductions to be made in the discharges of caesium-137, plutonium and other radionuclides', reported the Ministry. Which was nice. But, already by then, Windscale/Sellafield had been in operation for more than thirty years. Ten years later, during consultations on the drafting of revised authorisations for the discharge of radioactive wastes from the Sellafield site, the UK government noted that, 'Improvements have been made in the handling of discharged wastes, to diminish their impact. Discharges of radiologically significant wastes to sea from Sellafield are now running at about 1 per cent of the peak levels in the 1970s.' Such consultations and revised authorisations were necessary to enable the new Thermal Oxide Reprocessing Plant (THORP) to begin its reprocessing operations and to add to the level of historic discharges in the Irish Sea.

Today, the disposal of radioactive waste for licensed and non-licensed sites is regulated under the Radioactive Substances Act 1993 (RSA). Radioactive gaseous, liquid or solid waste may only be disposed of or moved off the site in accordance with official authorisations. According to the United Kingdom's Department of Trade and Industry, 'Authorisations for disposal of radioactive waste require the operator to use best practicable means to reduce discharges of radioactivity.' Operators are required to assess regularly all radioactive discharges, monitor the local environment, and report on the results. Failure to meet these requirements may result in the prosecution of the holder of an authorisation.

Scottish Worries

Although Sellafield is in England, and although the new Scottish parliament has limited powers over British nuclear policy, some MSPs have asked their colleagues in Edinburgh to note 'that the major source of radioactive pollution found on the Scottish coastline is not from Scottish nuclear installations, but from Sellafield in Cumbria', and 'that in testing for pollution in the sea off East Lothian, more man-made radioactivity can be attributed to Sellafield discharges than to the Torness nuclear generating station there'. Kenny

MacAskill and other MSPs say that, 'it is unacceptable that the entire length of the Scottish coastline should be polluted in this way by the BNFL plant in Cumbria'. Scottish and other MPs have also raised their concerns about Sellafield at Westminster.

However, at least one prominent Scotsman is well disposed towards the nuclear industry. Founder of the West Highlands Free Press, Brian Wilson is better known today as the United Kingdom's Minister for Energy and Construction with special responsibilities for nuclear policy. In his Scottish constituency stands the Hunterston nuclear power plant, which is owned by British Energy and which provides jobs for Wilson's constituents. He is thought to favour its replacement by a more modern nuclear reactor.

Reverse
Unfortunately, improvements in the official British attitude towards discharges to the Irish Sea received something of a reverse in the course of licensing Sellafield's new THORP and MOX plants. It was decided by British regulatory authorities that BNFL would continue to be permitted to dump radioactive technetium-99. That decision was a significant factor in persuading the Irish government to launch its international legal actions against Sellafield. Prior to permitting the continuing release of technetium-99 from Sellafield, the United Kingdom Environment Agency had engaged in a process of public consultation, which concluded in March 2001. Various governments and organisations expressed anxiety about the effects of allowing technetium-99 to be released.

During 2001, the authors of the contested 'WISE' report for the European Parliament observed that technetium-99 (half-life 213,000 years) discharges have led to particular concern. They claimed that, 'In 1997, technetium concentrations in crustacean – particularly in lobster – reached thirteen times the European Council Food Intervention Level (CFIL) in the vicinity of Sellafield.' They wrote that, 'Recent environmental surveys along the Norwegian coast indicate a six-fold increase in technetium concentrations in seaweed since 1996' and remarked how, 'In 1999, a number of high concentrations of various radio-nuclides were also recorded in fish, shellfish, sediments and aquatic plants, some exceeding CFILs several times.'

The Irish Consulted
Early in 2001, the Irish government made its opposition to such releases clear, through the Nuclear Safety Division of the Department of Transport. The United Kingdom had embarked on a consultation process, as required by regulations, and received a submission from Dublin which said that, 'The Irish Government wants to see an immediate cessation of radioactive discharges to the Irish Sea, including technetium-99 discharges. It is the view of the Irish Government that the Magnox stations should be closed immediately and that there should be an end to reprocessing.'

The Irish took the view that the English and Welsh Environment Agency ought to have used, as its principal assessment criterion, the commitment made by the United Kingdom government in its acceptance of the international 'OSPAR Strategy for Radioactive Substances'. This 'calls for the progressive and substantial reduction of discharges, which include technetium-99, with the ultimate aim of concentrations in the environment close to zero by 2020 for artificial radioactive substances'.

The Irish argued that the actual British mode of assessing Sellafield had 'an inherent bias toward disposal, i.e. the discharge of radioactive substances to the environment'. They claimed that, 'It appears to preclude consideration of options such as the continued storage of technetium-bearing wastes until systems are installed to eliminate their discharge or the cessation of the production of technetium-bearing wastes. The consultation document fails to address, in an adequate manner, the subject of justification for the technetium-99 discharges to the marine environment.' Referring to the fact that the safety of workers is mentioned in the consultation document, in the context of managing technetium-99, the Irish added that, 'While worker safety is very important, it must not be achieved by discharging radioactive substances to the marine environment.'

The Irish acknowledged, at the time of the public consultation, that Britain's general health and safety objective is the progressive reduction of the hazard of storing liquid radioactive wastes: 'The Irish Government strongly supports this principle, but not at the cost of discharging technetium-99 to the marine environment, with resultant radiation doses to man and biota.' The Irish argued that the need to protect the legitimate uses of the seas should not be undermined by arguments based upon concepts and cost-benefit analyses which use arbitrary monetary values for calculating detriment.

The Irish noted that a consultation document issued by the British themselves had stated that the current MAC (medium active concentrate) storage facility, B211, does not meet modern standards for nuclear chemical plant. The Irish said that, 'This is a startling admission. In our view, there should be a full public disclosure of the safety assessment of this facility and immediate steps taken to redress shortcomings.'

Nevertheless, the Irish noted that the UK Environment Agency had acknowledged that there is sufficient tank capacity in the B211 storage facility at Sellafield for existing and future MAC arisings to the year 2008, and perhaps longer: 'Tank capacity is therefore not a problem at present.' They proposed that this tank capacity be utilised to its maximum to produce an immediate substantial reduction in technetium-99 discharges and said that this was the most reasonable and acceptable option. They added further that, 'it should be noted that technetium-99 has a half-life of 213,000 years, and that the collective dose commitment has been estimated over 500 years for the European population. There is no general international acceptance that the collective dose calculation should be restricted to 500 years.'

The Irish believe that breaches of the European Union's Council Food International Level (CFIL), which sets limits upon the radioactivity in fish and other food, can only be assured if the alternative of storage is adopted: 'Otherwise, the contamination level will continue to exceed the CFIL.' They have criticised the English and Welsh Environment Agency's review of the options for the future regulation of technetium-99 discharges because it 'does not address, to any significant degree, the effects of discharges on the legitimate uses of the seas. Furthermore, public concerns are not addressed.'

Scottish support
The Scottish National Party is also worried about the impact of marine discharges from Sellafield. In January 2003, its leader John Swinney told me that,

> The SNP fully supports the Irish government in their attempts to stop the UK goverment polluting the Irish Sea with nuclear waste. As a party we have a long record of opposition to the pollution of the Irish Sea by Sellafield and have regularly registered our concerns about the pollution of Scotland and Ireland's coastal environment.

Swinney contrasted his party's concerns with what he says is the government's "complacency and disregard for the environment".

While dumping in the sea from Sellafield is better regulated than it was once, the government of the United Kingdom regards such dumping as an acceptable means of disposing of certain radioactive products and as being, sometimes, the lesser evil of two alternatives. This is so particularly where those dumped radioactive products result from any process that is deemed necessary to reduce the store of nuclear waste at Sellafield, and where the dangers to the public from dumping are thought by British officials to be far outweighed by the benefits to society which result from such a process.

CHAPTER 7

A 'Tiny' Risk?

The extent to which Sellafield poses a danger to people's health is hotly contested. Nobody denies that radiation can cause serious harm, but both BNFL and the government of the United Kingdom argue that the manner in which radioactive materials are handled and stored, and the way in which their security is protected, are the crucial factors in assessing the reality of any threat to the public's welfare. They say that their standards are high enough to prevent a disaster. Critics of Sellafield and of the nuclear industry, in general, query particular handling, storage and security procedures, while worrying also that the long life of so many nuclear wastes will make it virtually impossible to avoid an eventual catastrophe.

Not all nuclear materials emit radiation in the same way. Levels of radiation vary, as does the period during which particular materials remain dangerous. There are different forms of radiation, and radiation itself must be distinguished from contamination. Thus, an object that is not intrinsically radioactive may become contaminated by being exposed to radiation. However, it is not necessary to be a physicist to grasp the underlying reality, which is that nuclear fuels pose a threat to the population at large and must always be treated with great care. There is little reason to doubt that BNFL usually exercises such care. However, its standards occasionally lapse, and the possibility that human error or malice could fundamentally compromise the integrity of a facility such as Sellafield remains a live concern in the minds of many.

Natural and Unnatural Radiation

Radiation has existed since the beginning of the world. It is emitted by matter in the form of waves or particles. It can be natural or it can be created by people. Radiation from the sun is natural, for example. The X-rays used in medical examinations are generated by the human application of scientific knowledge. In normal circumstances, most radiation reaches us from natural sources, and not from the fallout from nuclear weapons tests or from accidents and releases at nuclear power plants. And not all manufactured radiation is unwelcome, as those who have benefited from certain medical and dental procedures will testify. Nevertheless, irradiation of body tissues may cause direct harm to the individual who is exposed, or harm future generations as a result of damage to human germ cells.

Radiation which is created by human activities is said to constitute less than 15 per cent of the total radiation that we absorb generally. And only a small proportion of even that amount of radiation is said to reach us from the activities of the nuclear industry. The nuclear industry claims that, in normal circumstances, it does not pose a major threat to the health of present or future generations.

When experts were assessing the extent to which people around Windscale/Sellafield had been exposed to radioactivity from the fire there in 1957, it was discovered that strontium-90 found on farms ten miles or more from Windscale was mainly from worldwide fallout from Russian and American nuclear weapons tests. The gradual accumulation of exposure due to radiation caused by man's activities, and the possibility of further catastrophic exposure due to war, terrorism or accident, remain causes of public concern.

Natural sources of radiation include outer space. Travelling by air increases the amount of radiation that we absorb from cosmic rays. BNFL has stated that even the Chernobyl accident increased radiation doses for a British individual, over his or her lifetime, by no more than an amount roughly equal to that received from cosmic radiation during a return air journey from London to New York. According to BNFL, 'at our Sellafield site, where reprocessing is the main business, the dose of radiation that a person living near Sellafield would receive in a year would be about the same dose as you would receive on a return flight to the Far East. This is because when you are flying, you are exposed to more naturally occurring radiation.' BNFL points out that an average person receives about twice as much radiation from flying for one hour as from the entire nuclear industry in one year.

Radiation also occurs naturally in the earth's crust and reaches us in the form of radon gas. The level of natural radon gas in particular areas may add to the risk of people who live there getting cancer. Advice is available on what parts of Britain and Ireland have high radon levels, and on the steps that may be taken by householders to counteract such levels.

Becquerels, Doses and Hype

When we eat, drink and breathe, we take in radioactivity that is present naturally in food and in the atmosphere. The emission of radioactivity is measured in tiny units, each of which is called a becquerel ('Bq', or 'MBq' in the case of one million becquerels). A much larger unit of measurement is the curie (ci). According to the United Kingdom's Department of Trade and Industry, one kilogram of coffee contains about 1,000 Bq of activity. Phew! And you wondered why you felt high after two mugs of the stuff?

In his recent authoritative book, *Permissible dose: a history of radiation protection in the twentieth century*, J. Samuel Walker traces some of the changing opinions on what constitutes a 'safe' dose of radiation, and also considers the significance of concepts such as 'low-level' exposure to radiation. The doses of radiation actually received by any particular individual are measured in units, each of which is usually known as a sievert (Sv). This is to be distinguished from the becquerel which, as we saw above, is the unit used to measure the radioactivity that an object emits.

Exasperated by 'hype' about the dangers from radiation, three physicists wrote to the *Irish Times* during 2002 'to put again the radiation from Sellafield into perspective'. Professor W. Walton of the National University of Ireland at Galway, Professor P.I. Mitchell of University College Dublin and Professor Ian McAuley of Trinity College Dublin pointed out that the current dose of radiation from Sellafield being received by the Irish public 'is largely obtained from eating fish', but that the dose from naturally occurring polonium-210 in the same fish 'is about 100 times greater'. The three scientists asked Irish people to compare the average 0.3 microsieverts of radiation from such fish 'with the 2,500 microsieverts per year we get from all natural resources' (*Irish Times*, 31 May 2002).

The United Kingdom's Department of Trade and Industry (DTI) draws an important distinction between two kinds of radiation, ionising and non-ionising. In a basic facts sheet, which it makes available in case of a nuclear emergency, the DTI notes that 'ionising radiation is important following a nuclear accident'. Ionising radiation can penetrate matter. According to the DTI, 'When people are exposed to ionising radiation tissues in the human body can become damaged, which may lead to harmful effects such as the development of cancer.' The level of harm arising depends on a number of variables, including the quantity of radiation that a person receives, the type of radiation and the period of exposure.

The time that it takes for radioactive material to lose half of its radioactivity by decay is expressed in terms of its 'half-life', and this is a common form of measurement. Some material has a half-life of days, including iodine-131, while other material may have a half-life of thousands of years.

What Effects are Measured?

Incidents at nuclear power plants may result in an escape of radioactive products that are then inhaled or ingested by members of the public. Once, assessments of the risks involved were considerably underestimated by official bodies. Slowly, a greater appreciation of the effects of exposure dawned. At the time of the Windscale fire of 1957, the effects of nuclear accidents tended to be measured simply in terms of whether or not a particular dosage of radiation had been absorbed by specific individuals in the vicinity of any radioactive spill or escape. Was there, or was there not, 'appreciable bodily injury' to any known person? Did their exposure pass a certain 'threshold'?

Subsequently, much importance has come to be attached to the relationship between an escape of radioactivity and the number of consequent deaths that are likely to occur in the population at large, even if no particular death can be proven to be causally connected to any particular escape of radioactivity. What is the extent of the 'collective dose'? For example, the fact that there was a low risk of any single person in Yorkshire suffering serious consequences from an inhalation of low-level radiation from the Windscale fire, in nearby Cumbria, did not alter the fact that, statistically, some people living far from Windscale might suffer consequences. Their particular illnesses or deaths resulting from the accident would be unlikely to be identified by their doctors as having been caused by Windscale and would be statistically insignificant relative to the number of other hereditary defects or deaths by cancer. Nevertheless, they would have been caused by it. Clearly, the worse the escape of radiation, the worse are any immediate and distant consequences.

Over the years, as scientific knowledge has improved, the level of risk from exposure to radioactivity has also been revised upwards. Exposure is now believed to be at least three to five times more dangerous than was thought to be the case at the time of the Windscale fire in 1957. The measure of effects has changed. Some would say that they are also unduly 'conservative', meaning restrictive.

Leukaemia?
In February 1990, the *British Medical Journal* published a study by Martin J. Gardner and others that appeared to establish a connection between the incidence of leukaemia near Sellafield and the cumulative dosage of radiation that had been received by workers at the plant itself. People in County Louth, across the Irish Sea from Sellafield, also worry that their children's health may have been affected by the plant.

The nuclear power industry is bound by law to take steps to protect its workers against excessive exposure to radiation. Since 1990, Gardner's study has been hotly debated. Other studies suggest that there may not be a clear correlation between the employment of parents in nuclear power plants and the likelihood of their children being blighted by leukaemia. It appears that the medical jury is still out on that particular issue.

In 1995, Chris Busby's *Wings of death* was published. In the book, he writes that ionising radiation from nuclear fallout and radioactive waste has been among the causes of various serious illnesses, including leukaemia, and is responsible for many premature deaths. Dr Busby, described as a physical chemist and member of the Green Party's regional council, representing Wales, includes 'a brief history of the radiation age and its consequences for human health'. Some of Dr Busby's more recent work, with the group 'Green Audit', argues that current risk estimates relating to Sellafield very seriously underestimate the real risk factor. Some of this work has been dismissed by Britain's official Committee on Medical Aspects of Radiation in the Environment (COMARE) as having been based on substantially erroneous data.

In 2001, the authors of the 'WISE' report for the European Parliament admitted that 'many uncertainties remain' when it comes to the relationship between nuclear power and public health. They wrote that,

The cause or causes of the observed increases in childhood leukaemia near Sellafield have not been determined, nor is it known whether a combination of factors is involved. The UK Committee on the Medical Aspects of Radiation in the Environment (COMARE) has stated: 'As exposure to radiation is one of these factors, the possibility cannot be excluded that unidentified pathways or mechanisms involving environmental radiation are implicated.'

Various hypotheses, including references to paternal preconception irradiation and population mixing, have been advanced without being conclusive. The 'WISE' consultants stated that, 'Possible explanations for the discrepancy between observed cancers and estimated low doses include erroneous dose assessments (in particular foetal doses) and uncertainties as to the parameter of "dose" and what it measures.' They stated that, besides childhood leukaemia, other causes for concern include reports of both an increased incidence of retinoblastoma in children and a statistically significant increase in stillbirth risk in the Sellafield region.

'Something as Emotive as Cancer'

In February 2002, two members of Sellafield's management team took part in a debate on Irish television in an attempt to quell fears about Sellafield emissions. The RTE programme began with opening summaries from Brian Wilson, United Kingdom Minister for Energy, and Dermot Ahern, an Irish government minister who also represents the people of County Louth in the Dáil. Louth is the closest county to Sellafield, and the people of Louth fear that their health has been adversely affected by the nuclear plant.

On the programme, John Clarke, Head of Environment, Safety, Health and Quality for the Sellafield site and Siân Beaty, Head of Safety for MOX, attempted to counter arguments about the impact of an accident at Sellafield. They

for the nuclear industry in Western Europe or in North America. A fire burned for several days and people gave their lives on site to bring the crisis under control. Tens of thousands of inhabitants were evacuated from surrounding areas, some of whom suffered severe long-term effects. Millions of acres of land will not be safe for agricultural use for decades or longer.

Some of the children of Chernobyl who suffered ill effects have been brought away on holidays by certain organisations in other countries. Irish people have responded generously to their needs, especially through the Chernobyl Children's Project, organised by Adi Roche and others (see www.chernobyl-ireland.com). Publicity surrounding the children who visit Ireland has heightened awareness of the dangers of a disaster at nuclear facilities closer to home, especially at Sellafield.

It is possible that other major or serious nuclear accidents have occurred internationally but have not been reported, particularly in countries where democracy and freedom of speech have not acted as a check against bad practice. However, it seems likely that evidence of such incidents would have been detected abroad, especially in more recent times.

The condition of nuclear power plants in Eastern Europe continues to be a cause for concern, particularly following the political, social and economic upheaval of the 1990s. A number of states have attempted to ensure that safety and security standards are maintained at such sites by contributing money and expertise for their upkeep and by concluding a number of international agreements. Further information on this situation may be found at http://www.dti.gov.uk/energy/nuclear/fsu and other places.

Major British Accidents
The United Kingdom's Department of Trade and Industry (DTI) sees little reason to worry about the possibility of major British nuclear accidents. According to the DTI, 'The

precautions taken in the design and construction of nuclear installations in the UK, and the high safety standards in their operation and maintenance, reduce to an extremely low level the risk of accidents which might affect the public.'

Those investigating a serious spill of radioactive liquid that occurred at Windscale, Sellafield, on 29 July 1978, found that 'a high-level alarm which might have given warning of a high level of liquid in the sump had been isolated, perhaps during a decontamination operation ten months before. An alternative level indicator had been taken off the regular monitoring schedule.' Some of the liquid spilled onto a roadway within the plant area (Charlesworth et al., p. 35). Such events are alarming, but the public fears far greater contamination from a major disaster.

Nuclear operators are required to prepare, in consultation with local authorities, the police and other bodies, emergency plans for the protection of the public and their workforce, including plans for dealing with an accidental release of radioactivity. These plans are said to be regularly tested in exercises under the supervision of the NII. The UK Department of Trade and Industry co-ordinates policy at national level as the leading government department in respect of the United Kingdom's arrangements for a response to any accident or emergency with off-site effects from a licensed civil nuclear site in England and Wales. Consequently, it chairs the Nuclear Emergency Planning Liaison Group, which brings together organisations with interests in off-site civil nuclear emergency planning.

In the event of an emergency at a civil nuclear site in Scotland, lead government department responsibility (and the main national co-ordinating role) would fall to the Scottish Assembly. The DTI would still be responsible for briefing parliament at Westminster and the UK's international partners. Further information on the planned British response to a nuclear emergency may be found at http://www.dti.gov.uk/nid/er_factsheet.htm.

The Earth Moved

In the autumn of 2002, an earthquake registering 4.8 on the Richter scale struck the English Midlands. Buildings shook, chimneys fell and masonry came tumbling down. The worst British quake in a decade, it was a reminder that tremors do occur regularly in England and raised questions in people's minds about the ability of storage tanks and other facilities to withstand the shock of any stronger than usual English earthquake. BNFL believes that such quakes pose no threat to the safety of its site at Sellafield and say that they have been taken into account as an eventuality.

What if?

Opinions differ when it comes to assessing the effects of a possible major event at Sellafield. Take, for example, two articles in the *Irish Times*. On 5 October 2001, Dick Ahlstrom, that newspaper's esteemed science editor, played down the likelihood of dangers from terrorism or accident and attempted to focus instead on the detrimental environmental aspects of Sellafield. He wrote that activities on the site

> take place in heavily reinforced buildings and so a significant impact would be required to breach them. If this happened and aviation fuel caught fire, then there is potential for radioactive material to be carried upwards. Large stores of exposed radioactive material would need to be available to make this a significant threat, as at Chernobyl where an explosion ripped open the core, exposing tonnes of fuel to the fire that followed.

Ahlstrom concluded that,

> Two of the worst possibilities are the rupture of the high-level liquid storage tanks or a fire involving plutonium. The latter is a dangerous substance even at low levels. A strong fire has the potential to spread

radioactivity, but it would have to be a large fire and the wind would have to be blowing in our direction to cause us harm.

But just a few weeks later, on 10 November 2001, Nuala Ahern, a Green Party Member of the European Parliament, passionately argued that there is an imminent danger to the Irish public from Britain's nuclear activities:

A terrorist attack on Sellafield would be devastating to Ireland. The worst scenario would be an assault on Sellafield's dangerous high-level waste tanks. These are full of volatile radionuclides, such as caesium-137. Recently, BNFL closed the THORP plutonium reprocessing plant because the high-level waste tanks were unable to cope with the current level of waste. Radioactivity spread from an attack on Sellafield could render large areas of Ireland uninhabitable, and agriculture ruined forever.

In *The Engineers Journal* of September 2002, Frank Turvey recalled a public forum on Sellafield that was held at Drogheda, Co. Louth, in November 2001. Turvey, a member of the board of the Radiological Protection Institute of Ireland, stated that his institute's then chief executive told the forum that 'a severe accident involving the high-level waste storage tanks would give rise to about thirty fatal cancers in an expected population of 50,000'. He reportedly added that there would be no immediate deaths. There would, he thought, 'be considerable damage to our fishing, tourist and agricultural industries'. If that seems almost reassuring when set against the warnings of Nuala Ahern and other opponents of Sellafield, the projections of a prominent United States expert on nuclear power are alarming. In comments published by the Royal Irish Academy during 2002, Dr Richard Garwin remarks that storage tanks at Sellafield contain thirty times as much caesium-137 as did the reactor at Chernobyl. He believes that 'the consequences would be

CHAPTER 8

Ships That Pass

A t Barrow-in-Furness, Captain Kerry Young welcomes us on board his ship, the *Pacific Crane*. The vessel brings spent nuclear fuel from Japan to Sellafield. Two BNFL subsidiaries, Direct Rail Services Ltd (DRS) and Pacific Nuclear Transport Ltd (PNTL), are employed to carry nuclear waste by rail and ocean, and to export reprocessed fuel in the same manner. Cogema of France and the Japanese nuclear power utilities are minority shareholders in PNTL.

BNFL is satisfied that its transport systems are safe. Those who live along the coasts of Ireland, England, Scotland and Wales hope so. In a reception room at Barrow-in-Furness, under the dignified gaze of an official portrait of Her Majesty, Queen Elizabeth II, Captain Malcolm Miller boasts to me that 'BNFL has the safety record to end all safety records.' Miller is BNFL's Head of Operations Transport and he points out that PNTL's purpose-built ships are classified as the highest safety category of the International Maritime Organisation. The chances of the Irish Sea becoming one big radioactive cooling pond are slim. Really!

PNTL ships dock at the company's terminal in Barrow-in-Furness, just down the Cumbrian coast from Sellafield. The company has five ships for sailing to Japan, a couple of which have been adapted especially to deal with the heightened security that is required for transporting MOX fuel because of its particular plutonium content. These are the *Pacific Pintail* and the *Pacific Teal*. The *Pacific Crane* is deployed on runs to Japan but is not one of the vessels adapted for carrying MOX fuel. It currently lies empty, awaiting its next voyage abroad. It is, for example, regularly despatched to collect spent fuel for reprocessing. The fact that it is lying idle provides an opportunity for me to see the vacant holds in which such cargoes of spent nuclear fuel are usually kept.

Standing with me on his bridge, as we gaze across a variety of control panels and dials out towards the Irish Sea, Captain Young stresses that his ship is built with safety in mind. There is a duplicate or spare of everything that is deemed essential. There are two propellers, and two rudders in line with each propeller, for example. There is also a propeller in the bow ('bow thrust') to make the vessel more manoeuvrable. The ship can be turned around within its own length. There are two engines. There are also two powerful radars, for picking up approaching ships from over twenty miles away. The adapted MOX ships also have special thermal imaging devices to pick up on their screens even rubber dinghies. The five ships used for the Japanese runs, including the *Pacific Crane*, are each roughly equal in size. They are about 105 metres long, which is a factor determined by the size of the Japanese reactor ports which they serve.

Captain Young is eager to point out that even in a disaster, if an absolutely huge ship were to hit the *Pacific Crane* or other members of the fleet at high speed and plough through a bulkhead, the PNTL vessels are built to stay afloat. Each has a double hull and watertight doors that can seal off sections of the vessel from one another. Compartments of PNTL ships can also be deliberately flooded, if required. This may be necessary to protect nuclear cargoes in certain circumstances. Any water that happens to become radiated in the process gets held in a tank and is monitored. If the radiation levels are found acceptable to BNFL and its regulators, the water may then go over the side and into the ocean.

Clean Room

We clamber down into the bowels of the ship, entering 'The Clean Room' where all monitoring equipment is kept. From here, once a week, a thorough check of the whole vessel is undertaken. Behind 'The Clean Room' is a door into the restricted area, where only six designated officers are permitted during voyages. Through this door, we descend into the hold below No. 5 Hatch. It is an airless, featureless space, surmounted by a giant hatch that can be lifted off by crane to allow a flask of radioactive fuel to be lowered and bolted to the floor, for transporting across the ocean. The flask, says Captain Young, effectively becomes part of the ship's structure when locked in place. He points to the walls and says that cables are duplicated on each side of the vessel, allowing a tremendous reserve of power for all eventualities.

We also inspect the engine room of the *Pacific Crane*. Again, certain of the ship's double abilities are pointed out, including its two gearboxes. Even on a mild English summer's day, with the vessel stationary, the engine room is a very warm place. What it must be like at certain points on the journey to or from Japan can only be imagined. The crew in here no doubt gets very thirsty on a long, hot voyage, but they are not allowed to indulge unduly in alcohol when they get off duty. The distribution and consumption of alcoholic beverages on board is strictly controlled.

On the PNTL dockside in Barrow stands a massive green crane, awaiting the next time that it will be required to lift gingerly a flask of radioactive fuel in or out of one of these ships' holds. Also docked at Barrow, facing the *Pacific Crane*, sits the *European Shearwater*. This is one of two PNTL ships that ferry fuel between Sellafield and Europe. Until the mid-1990s spent fuel from Europe used to go through ports along the English Channel, on roll-on, roll-off ferries.

The Package Matters

According to Captain Malcolm Miller, the conventional wisdom in relation to nuclear transports is that, 'The conveyance does not matter. The safety depends on the package you are using.' Nuclear fuel is transported within specially constructed containers known as flasks. These are each bigger than a family car. They are made of stainless steel, or steel and lead. They have walls one-foot thick and weigh up to one hundred and ten tonnes when full, this weight consisting of more than 90 per cent protection and less than 10 per cent fuel. Each such package costs about £1 million sterling. Vivid demonstrations have been mounted in the past to show how strong these flasks are. A heavy railway locomotive with three carriages, travelling at a speed of 100 miles per hour, was deliberately crashed into one such flask placed on a railway line. It is said that the locomotive was destroyed while the flask remained intact.

BNFL asserts that, 'The safety record for transporting nuclear materials is second to none. Since the early 1960s, nuclear material has been transported over 16 million miles by rail, road and sea without a single accident resulting in a harmful release to the environment.' PNTL boasts of having transported more than 4,000 flasks of nuclear fuel and covered 4.5 million miles without a single incident involving an increase in levels of radioactivity.

Plutonium, about which there is particular public concern, is usually transported as oxide powder or as mixed oxide (MOX) fuel rods by air, land or sea, in specially designed containers. BNFL says that,

> For transporting plutonium powder by air, the containers have been shown to stand up to damage equivalent to a fall from an aircraft at high altitude on to a hard surface such as frozen ground or concrete. For transporting the powder by sea, the ships used are designed to the highest possible standards with the most up-to-date safety equipment. The containers themselves are designed to stand up to the most fierce fire possible on board ship and pressure tested for being dropped in water.

One shipment of MOX fuel is said to carry the equivalent energy capacity of twenty-five tankers transporting 200,000 tonnes of oil each. Such tankers are more likely to cause a great deal of pollution than any PNTL vessel, in normal circumstances.

PNTL brings waste for reprocessing not only to Sellafield but also to the French reprocessing plant at Cap de la Hague. In 2001, its shipments included a consignment of MOX fuel sent to Japan from France. In 2002, it included a shipment of MOX fuel sent back to England from dissatisfied Japanese customers. Shipments of fuel to and from Japan have taken place via the Panama Canal. However, because shipments of MOX fuel must avoid any 'choke point', these particular loads are always taken to Japan the long way, round the Cape of Good Hope in Africa or Cape Horn at the tip of South America.

Captain Malcolm Miller says that the standards of PNTL are far higher than those required by international regulatory bodies. He adds that a normal licensing requirement is that a flask containing radioactive materials must be capable of withstanding pressure at 200 metres below sea level. However, he notes, 'we did a study of our packages and they would not be causing a problem until 8–10,000 metres'. He also reports that the best salvage companies claim to be able to retrieve packages from any level, provided that someone is prepared to pay the bill. And, even if a MOX pellet could somehow find its way out of its protective flask and into the sea, it would take thousands and thousands of years to dissolve, according to Captain Miller.

A Big Gun

The vessels that carry MOX fuel through the Irish Sea and beyond are fitted with a big gun, among other special defence items. But BNFL is extremely coy about precisely what protection is available to the captain of any of its ships. Those carrying MOX fuel have extra accommodation and gyms for armed police officers, who do not normally travel with shipments of spent fuel being sent to Sellafield and who are clearly expected to keep fit.

The movements of all PNTL ships are monitored around the clock from a report centre in England. Every two hours a special signal is sent by each ship. In the event of a signal not being received on time, an emergency plan is put into operation. Each vessel also downloads an electronic 'snail track' onto a computer in England. It is also equipped with the Global Maritime Distress Signal, allowing it to send out a signal that will be picked up internationally.

Asked about the possibility of an attack at sea by a well-organised terrorist group, Captain Miller is reluctant to go into details of his company's contingency plans. He does say that one of the reasons that PNTL ships have never gone to Japan via Suez (although some of them not carrying MOX do pass through the Panama Canal), is that the route would bring them through seas in the East that are renowned for piracy. BNFL points out that even if a ship were to be boarded, it would be no easy thing to access its nuclear load, and the ship could not be taken anywhere without it being noticed. So might a hijacked ship be deliberately sunk by its own crew, or even by armed forces if necessary, with a view to retrieving its cargo later from the seabed? I do not receive a clear answer to this query.

More Big Guns
The Secretary of State for Transport accounts to parliament for the safe transport of nuclear materials by road in Britain. This eminent individual sets the parameters within which BNFL is responsible for safely moving nuclear fuels and waste. The director of the Office for Civil Nuclear Security (OCNS) also regulates certain arrangements by the civil nuclear operators for the secure transport of sensitive categories of nuclear material. Thus, the OCNS is the United Kingdom's designated national authority under the Convention on the Physical Protection of Nuclear Material, for shipments to and from overseas destinations. Under the

regulatory control of the OCNS, for example, fall the arrangements for the shipments of MOX fuel to Japan. The first shipment took place in 1999 and the second, of French-manufactured MOX fuel, between January and March 2001, in British flagged ships operated by PNTL. According to the OCNS, both of these PNTL ships carried deck-mounted naval guns and an armed escort provided by the UKAEA Constabulary: 'Other security measures were also taken to prevent boarding or unauthorised interference with the MOX fuel.'

The OCNS notes that former bans on the transport of irradiated fuel imposed by Germany and France, due to the presence of low levels of contamination on the flasks in use, were lifted after approvals from the safety authorities. Shipments from countries on the European mainland to Sellafield for reprocessing resumed in late 2001. 'In addition', reported the OCNS during 2002, 'over 400 transports of less sensitive fissile nuclear materials, including irradiated nuclear fuel, have been carried out within the UK and to and from overseas destinations' between April 2001 and March 2002: 'This number includes regular movements by train of spent fuel from the nuclear power stations sent to Sellafield for reprocessing. There were no security incidents associated with any nuclear transports.'

S.O.S.
On the bridge of the *Pacific Crane*, Captain Young recalls an incident during one voyage when he diverted his vessel in response to a distress signal and rescued the crew of a cargo ship that was in flames. He says that there are special security procedures in place to process people coming on board, and to bring them to a nearby port, in such circumstances. The BNFL ship itself is not obliged to dock, but may allow the rescued personnel to disembark to a small boat. Young's action in diverting to save the crew of a burning ship was both humanitarian and in accordance with maritime practice. However, in the jaundiced aftermath of the attacks of 11

September 2001, one wonders if pirates or terrorists could use such a fire to lure a ship carrying nuclear fuels into a trap.

Passing the Buck

Down the years, the regulations governing the transportation of nuclear materials have left a lot to be desired, and demonstrate that the industry cannot necessarily be left to manage its own affairs adequately. According to the director of the OCNS, writing in his report published in June 2002, 'At present, nuclear operators are required to ensure that their contracts with carriers require the latter to take adequate security measures. However, it is difficult for operators to ensure that effective security is maintained throughout the journey and there is a strong case for placing responsibilities on those best placed to discharge them.' He went on to point out that, under proposed new regulations, those intending to transport sensitive nuclear material would henceforth need to be approved by the OCNS, by submitting a statement of their security systems and arrangements, such as is required already for site operators: 'Given the critical need to ensure the security of nuclear material while in transit and the concerns that arise from time to time about the protection currently provided, this is a necessary extension of regulation in the national interest.' It does seem remarkable that it appears to have taken the attacks of 11 September 2001 to awaken some people in the nuclear industry to the need for adequate security.

International Rules

In 1994, the Irradiated Nuclear Fuel (INF) Code was developed under the auspices of the International Atomic Energy Agency (IAEA), the International Maritime Organisation (IMO) and the United Nations Environment Programme (UNEP). The INF Code sets out how nuclear material, including MOX fuel, should be carried. Also in the 1990s, a working group composed of representatives from the IAEA, IMO and UNEP commented on the safety of transport of radioactive materials and the adequacy of the IAEA regulations as follows: 'All the available information demonstrates very low levels of radiological risk and environmental consequences from the marine transport of radioactive material . . . It was the unanimous conclusion of the Member States that there was no information or data that would cast doubt on the adequacy of the IAEA regulations.' This working group also established a major co-ordinated research project on the severity of accidents in the maritime transport of radioactive material. The researchers concluded 'that the risks of maritime transport in type B packages [i.e. the packages used for transporting highly radioactive materials] of highly radioactive material are very small.' All transports by the United Kingdom are said to be undertaken in full compliance with all relevant standards.

Irish Unimpressed

The Irish government has made it clear to the International Tribunal for the Law of the Sea that it is unimpressed by BNFL's argument that its ships are safe. In the autumn of 2001, the Irish complained to the tribunal that the probabilities of collision and fire on board the MOX carriers have been assessed only for the 'at sea' legs of the voyage, excluding the risks of collisions, rammings, groundings, fire, explosions and foundering when the carrier vessels are in the approaches to ports and berthing in harbours. They claimed that, 'A fierce fire could cause the plutonium in the MOX fuel to vaporise resulting in the release of a large number of respirable particles into the atmosphere and the marine environment. If these were to be blown over land it would amount to a serious hazard to the population.' They noted that if the ship were to sink, any unrecovered containers of nuclear fuel would eventually corrode and release MOX fuel into the sea.

The Irish are also concerned about the consequences of a terrorist attack on a MOX ship. They fear that terrorists could seek to take MOX fuel from a ship and to separate the plutonium from the MOX fuel to produce a nuclear weapon. They pointed out to the International Tribunal for the Law of the Sea that, in 1999, the chairman of the US House of Representatives International Relations Committee wrote to the then Secretary of State, Madeleine Albright, expressing concern about MOX deliveries to Japan by ship. He stated that 'with a top speed of 13 knots [the ships] would not appear to have sufficient defensive and deterrent ability much less the manoeuvrability or speed of military or coast guard escort ships'. The Irish added that, according to *Janes*, a periodical which is a recognised authority on military and naval matters, the ships are 'capable of repelling only a light armed attack' and need to be protected by 'at least one well-armed frigate'.

That MOX Return

While two shipments of MOX fuel were successfully transported to Japan between 1999 and 2001, the contents of the first shipment were certainly not satisfactory. That fuel was rejected by the Japanese utilities after BNFL had revealed that quality measurement data had been falsified by members of its staff. The fuel was held for return to Britain in 2002.

Prompted by this imminent shipment, the OCNS reviewed security measures in the context of the terrorist attacks in the United States, as did regulatory authorities in the United States and Japan. A few weeks before the ship left Japan, the director of the OCNS commented, 'All the authorities concerned are satisfied that the security arrangements to be taken are amply robust to deal with any potential threats. My office has taken note that some degree of interest in this shipment may be anticipated from anti-nuclear groups.' BNFL announced that, for this particular voyage, the fuel assemblies were being transported in Excellox 4 MOX transport casks which had been built to, and certified to meet, the exacting standards required by the International Atomic Energy Agency (IAEA): 'The underlying principle of these standards is that transport safety is ensured by secure packaging.'

BNFL noted that the forged steel transport casks weighed between 80 and 100 tonnes and had been specifically tested for security and reliability against the requirements of the International Atomic Energy Agency, including the following:

- Drop tests, during which the lid seal must remain intact after being dropped one metre onto a concrete and steel reinforced spike, and a nine-metre drop onto an unyielding surface – all performed at angles which ensure the maximum impact on the cask.
- Fire testing, requiring the cask to withstand an all-engulfing fire of 800 degrees Celsius for 30 minutes – a test far more destructive than any real fire aboard a vessel.
- Pressure tests, in which the cask must withstand pressure of at least 15 metres of water – in fact the casks are able to withstand the pressures created by submersion in several thousand metres of water.

BNFL added that the ships themselves were classified as the highest safety category of the International Maritime Organisation, and employed numerous safety features. These included the double hulls that were mentioned above, 'effectively making them a ship within a ship'. BNFL added that 'the vessels were designed to withstand a severe collision with a much larger vessel without penetration of the inner hull'. They are also said to have enhanced buoyancy, enabling them to stay afloat in extreme circumstances. They have a British crew that BNFL says is twice as large as crews on chemical tankers of a similar size, and all engineering and navigation officers hold higher than usual qualifications.

Sinking
According to BNFL, a study by the Japanese Central Research Institute of the Electric Power Industry shows that even in the highly implausible circumstances of a vessel sinking and the cask with MOX fuel being breached in coastal waters, 'the impact on those living near the incident would amount to

one-millionth of natural background radiation. If such an incident occurred in deep waters, the impact would be equivalent to one ten-millionth of background radiation.' BNFL says that a fully trained and equipped team of marine and nuclear experts is available on a 24-hour emergency standby system, required by the International Atomic Energy Agency, to deal with the unlikely event of a vessel getting into difficulty. In the unlikely event of a vessel being lost at sea, it can be located by the sonar location system built into the ship and said to be capable of operation to depths of over 6,000 metres. This sonar system has a range of twenty kilometres and is intended to relay to the surface the depth and angle of the vessel, whether the vessel is distorted or broken, whether the hatch covers are in place, details of the radiation level in each hold, and the temperature. A contingency salvage plan, utilising a salvage group with worldwide capabilities, is in place.

Nevertheless, the Irish government remains unconvinced and regards MOX shipments as an unacceptable risk to the environment of Ireland and to the health and economic well-being of its population. As the *Pacific Pintail* and the *Pacific Teal* approached Britain from Japan in the autumn of 2002, with their cargo of returned MOX fuel, Ireland's Minister for Communications, Marine and Natural Resources, Dermot Ahern, objected to the Irish Sea being used as what he described as a 'nuclear fuel highway – the final destination for other nations' nuclear waste'.

CHAPTER 9

Terror and Theft

The Scottish and Welsh nationalist parties were angry. They felt that the people whom they represent were not adequately protected in the aftermath of the terrorist attacks in the United States on 11 September 2001. At the end of November 2001, at a joint news briefing in the House of Commons in Westminster, Scottish National Party and Plaid Cymru speakers did not pull their punches. The SNP's Alex Salmond said that,

> Two weeks ago, the International Atomic Energy Agency convened a special conference in Geneva in order to discuss the terrorist threat to nuclear installations. The Director General of the IAEA, Mohamed El-Bareida, warned: 'Nuclear materials can be stolen and nuclear facilities are vulnerable to attack. There is no sanctuary any more.' And nuclear scientists at the conference said that the British government is doing 'woefully little' to protect its nuclear power plants and storage facilities. At the Sellafield plant in Cumbria, an estimated 2,000 cubic metres of waste is held in open ponds. And one of Britain's leading nuclear scientists, Dr John Large, has said: 'If terrorists hit those ponds it would be devastating . . . The threat of turning Sellafield into a dirty bomb is real and we really do not have any defence against aerial attack'

(*Sunday Mail*, 11 November 2001). Dr Large has identified six possible targets for terrorists at Sellafield – including tanks holding huge amounts of highly radioactive caesium-137.

The possibility of an aircraft crashing at Sellafield had been considered by the British authorities long before Manhattan's Twin Towers were hit. However, the authorities can scarcely have imagined acts as deliberate and devastating as those which had made leading politicians in Scotland and Wales so agitated.

A 'Remote' Likelihood

Years before the terrible attacks on New York in 2001, the possibility of an aircraft hitting a nuclear power plant was considered by the British. Certain standard 'Safety Assessment Principles for Nuclear Plants' adopted by the United Kingdom's Health and Safety Executive had advised that, 'The predicted frequency of aircraft and helicopter crashes on or near safety-related plant at the nuclear site should be determined' and that the risk associated with the impacts, 'including the possibility of aircraft fuel ignition', should also be taken into account. The ultimate test to be applied was whether or not it could be demonstrated 'that the frequency of an event being exceeded is less than once in ten million years, or if the source of the hazard is sufficiently distant that it cannot reasonably be expected to affect the plant'. These Health and Safety Executive principles, adopted prior to 11 September 2001, added that,

> the calculation of crash frequency should include the most recent crash statistics, flight paths and flight movements for all types of aircraft and take into account forecast changes in these factors if they affect the risk. Relevant bodies should be consulted by the licensee [BNFL in the case of Sellafield] with the object of minimising the risk from aircraft approaching or overflying the plant.

In February 2001, referring to facility B215 at Sellafield, in which there is stored a large quantity of highly radioactive and dangerous liquids, the NII of the Health and Safety Executive admitted that, 'There has been no specific design provision against crashing aircraft.' However, they reported that, 'The effects of aircraft crashes have been assessed and BNFL concludes that in absolute terms the likelihood of an aircraft impact onto any individual plant is very remote (i.e. the total impact is below 1×10^{-6}/year).' But old computations of 'likelihood' based on accident rates and normal aircraft cruising speeds are inadequate in an age when hijackers choose to aim aircraft deliberately at buildings and to fly at speeds far in excess of those that are usual during civilian flights.

In 2002, when I visited THORP, there appeared to be nothing more than a thin layer of roofing between the storage ponds and open sky. And while the roofing may offer some protection against errant light aircraft, limited 'no-fly' zones around Sellafield are unlikely to make much difference in the event of a determined approach from the skies at high speed. At that point, the protection of millions of people in Britain and Ireland would depend largely on the alertness of air traffic controllers and the response capability of the Royal Air Force, as well as the sturdiness of buildings and storage tanks at Sellafield.

Not all air crashes at Sellafield would necessarily have severe radiological consequences. Some might be extremely nasty for the Sellafield workforce, without there being a significant release of radioactivity far beyond the perimeter of the plant. However, no clear scientific calculations of the consequences of an attack on Sellafield similar to that on Manhattan's Twin Towers have been published. Quite detailed maps of Sellafield *are* available on the web.

11 September and Costs
The events of 11 September 2001 in New York and Washington were, apart from anything else, a vivid

illustration of the vulnerability of key installations to aerial attack. In Britain and throughout the rest of the world, governments and regulatory authorities immediately began to reassess their levels of security.

On 10 October 2001, Joe Jacob, the Irish junior minister responsible for matters nuclear, told senators in Dublin that the events in New York the previous month 'show that there is a huge onus on those countries with nuclear installations to protect such installations from attack. Those countries have, in my view, an absolute duty to do everything possible to protect their dangerous installations from attack, irrespective of the cost of doing so.'

What if someone seized control of an aircraft and flew it into a nuclear plant such as Sellafield? The director of the UK's Office for Civil Nuclear Security (OCNS) admitted in a report published by him in June 2002 that, 'These ruthless, indiscriminate, but well-planned and executed, attacks prompted a significant reassessment of the threat posed by Islamic extremist terrorist groups.' Quite why he singled out Islamic groups is not clear, although it will become evident below that the Irish were also on his mind. Is Sellafield not vulnerable also to aerial assault by a group of crazed Anglicans from Kent?

In his report, the director sought to be more reassuring than illuminating about the measures taken to protect the public against the consequences of such an attack on a nuclear facility. He wrote,

> Although the details must remain confidential, I can confirm that my Office undertook an immediate, thorough review of the implications for the civil nuclear industry, based on advice provided by the Security Service. We also worked closely with the NII [Nuclear Installations Inspectorate] undertaking several joint audits on the ground at sites.

The director noted that,

Although public attention since those attacks has focused almost exclusively on the danger of another attack using hijacked passenger aircraft, we had to take account of the possibility that any further attacks by Al Qaida, or any other extremist Islamic terrorist organisations, might be mounted from the ground.

He added that terrorists would assume, correctly, that precautions against hijacking would now be much more stringent in the wake of those attacks and noted,

We already had in place comprehensive, stringent security arrangements, principally against the threat posed by Irish republican terrorist groups. Nevertheless, chicanes and other measures were put in place promptly or strengthened around all civil nuclear sites, in case terrorists sought to use vehicles loaded with explosives to crash through perimeter defences. Security arrangements protecting certain sensitive areas inside sites have also been extended and reinforced.

Anti-Aircraft Guns
Following the attacks of 11 September, there were calls for the installation of anti-aircraft guns at nuclear sites. The French placed surface-to-air missiles around their nuclear facility at Cap de la Hague. Restrictions on flying over nuclear sites seem somewhat inadequate, especially taking into account the speed at which jet airliners fly and the fact that those which hit the Twin Towers in New York were flying even faster than is usual for civilian aircraft. The possibility of Royal Air Force fighters being scrambled in time to intercept and shoot down a jet before it plunged into Sellafield, or any other nuclear facility, is less than certain. However, in his report of 2002, the director of the OCNS could not resist a dismissive comment on the French measures, which do indeed appear to have been largely for show. He wrote that,

measures have been taken to protect civil nuclear sites and Sellafield in particular from the possibility of any further attacks using hijacked aircraft, in conjunction with the Ministry of Defence. Although various options were given very careful consideration, the measures selected involved strengthened warning procedures and interdiction by RAF interceptor aircraft. Commentators in the media made much of a decision by the French authorities, following the attacks in the United States, to station surface-to-air missiles around the large nuclear site at Cap de la Hague, near Cherbourg, as well as at other, selected non-nuclear sites. However, national circumstances differ. Moreover, the missiles around Cap de la Hague have since been removed.

Britain's nuclear facilities are also protected by a network of specialist committees and working groups established under the auspices of the UK Cabinet Office to review security and emergency planning arrangements.

Top Secret
Critics remain less than fully convinced by the reassurances of British authorities that the chances of a successful air assault on Sellafield are slim, or by predictions that it is unlikely that an aircraft crashing into facilities there would have catastrophic consequences beyond the locality. However, questions about the details of security protection at the plant are met by a stock response that all eventualities have been considered, and a hint that there is much more than meets the eye to the protection that has been put in place against any awful eventuality. One hopes that there is.

The ability of British authorities to prevent an effective terrorist strike on a nuclear power plant remains uncertain, partly because those same authorities feel so constrained not to provide information that might be of use to anyone plotting an attack. The need for security means that inefficiencies are more difficult to discern than would otherwise be the case. As the director of the OCNS put it in his own report of 2002,

> In the immediate aftermath of the attacks in the United States, many questions were asked about the vulnerabilities of civil nuclear sites and security measures in place against terrorist attacks. In the interests of national security and in keeping with long-standing Government policy, very little could be offered in reply, except assurances that security is kept under continuous review and stringent measures are being taken against the risk of terrorist attack.

He acknowledged that the nuclear industry has often been criticised for sheltering behind a cult of secrecy inherited from its origins in the government's nuclear weapons programme. He observed that in response to such criticism (and to meet various statutory requirements on disclosure) a certain amount of information has been published in recent years.

While cautioning that there is a danger that terrorist groups might exploit publicly available information collated from a variety of sources, including the Internet, he said that an expert group composed of representatives from the main operating companies and the industry's regulators has been reconsidering the balance between providing information of legitimate public interest and protecting the national interest against terrorism and proliferation.

The director concluded his report on a note of confidence. He asserted that 'satisfactory arrangements' had been made with the Ministry of Defence to protect civil nuclear sites from attacks from the air. The public will hope that his confidence is well founded.

One must also hope that sufficient facilities have been installed at Sellafield to respond adequately if a commandeered jet does get through and crashes onto the site or if an employee engages in sabotage. Quick intervention might prevent the levels of cooling water dropping below a critical level for a critical period, but such intervention depends in the first instance on the design and installation of necessary equipment, including remotely operated cranes and special sheeting and emergency fuel-cooling water.

Trusting the OCNS

The OCNS, which has been part of the United Kingdom's Department of Trade and Industry since October 2000, is the security regulator for the United Kingdom's civil nuclear industry. Administered as an independent component of the Nuclear Industries Directorate, it is responsible for setting security standards for the industry and enforcing compliance. Through the Standing Committee on Police Establishments (SCOPE), the OCNS also reviews police numbers at licensed nuclear sites policed by the United Kingdom Atomic Energy Agency (UKAEA) Constabulary.

Prior to October 2000, the nuclear industry appears to have dragged its collective heels on security matters, exposing the public to a greater risk than might otherwise have been the case. In his report of June 2002, the director of the OCNS remarked that, 'The transfer of OCNS to the Department has significantly improved the effectiveness of this office in regulating civil nuclear security. Hitherto, its anomalous position had inhibited the responsible authorities from providing a full service of intelligence reporting. These difficulties disappeared virtually at a stroke, following the transfer.' He stated that, in the past, the operating companies had been 'inclined to prevaricate in complying with security standards and guidance issued by this office from within the UKAEA, or to appeal to the Department's policy and sponsor officials', which created further delays.

The reformation of the OCNS as an independent and fully autonomous unit of the Department of Trade and Industry followed an official inquiry into various safety and security issues at the nuclear facility at Dounreay. According to the present director of the OCNS, writing in his report of 2002,

'It had become increasingly untenable that the Government's security regulator should be legally a component of a nuclear operator itself subject to regulation, contrary to guidelines issued by the International Atomic Energy Agency.'

The mission statement of the OCNS is as follows:

> To ensure that the nuclear materials and sensitive information of the civil nuclear industry, and those employed in the industry, are effectively protected against deliberate acts that threaten national security, the environment or public safety, and help retain public confidence, without imposing unjustifiable burdens on the companies subject to regulation.

The OCNS employs approximately three dozen people, with annual expenditure of around £1.6 million sterling. Most of its security specialists are recruited from outside the Department of Trade and Industry, after careers in the security and intelligence agencies, the armed forces and the police. However, applications are also welcomed from industry specialists, or graduates with degrees in risk and security management. Demand from the private sector for experienced security professionals has grown in recent years, and the director of the OCNS has problems retaining some of his best staff. In such circumstances, given the vital nature of the job that OCNS does, it is disturbing to read that the director lost two experienced inspectors in the eighteen months up to March 2002 and that he faced considerable difficulty and delay in recruiting replacements. More of his experienced staff have retired or left since then, compounding the difficulties of finding suitably qualified replacements and filling new posts. There is the added strain of helping to provide advice and assistance internationally, in Russia and Eastern Europe, for example.

Ultimate Security
The job of making sure that nuclear facilities are secure is, in the first instance and ultimately, a matter for nuclear

operating companies such as BNFL. It is not primarily the responsibility of official bodies such as the OCNS. The guiding principle for security arrangements at BNFL plants is 'defence in depth', meaning that they must never rely on just a single precautionary measure in any case.

Because the potential threats to be countered at nuclear power plants could affect national security, governments provide specialised support from the security and intelligence agencies, the police and the armed services. In order to retain public confidence, the United Kingdom government sets out to demonstrate 'that security requirements cannot be cut back or undermined by commercial pressures, although the companies themselves are entitled to expect that the security standards and procedures imposed are reasonable and realistic'.

Theft
Early breaches of security relating to nuclear matters involved spies, such as Klaus Fuchs, convicted of passing atomic secrets to the former Soviet Union. Espionage is still a concern of the American and European authorities. With the end of the Cold War, they see the main threat to security coming from countries such as Iraq and Iran attempting to circumvent export controls to acquire proliferation-sensitive technology to further their own nuclear weapons programmes. According to the director of the OCNS, 'The proliferation threat is often overlooked by commentators but the dangers are real, as illustrated by the conviction of an individual in Germany in 2000 for passing details about sensitive uranium-enrichment technology to the Iraqi authorities.' Since the late 1960s, the threat of terrorists attempting to sabotage a nuclear facility, or stealing fissile or radioactive material to fabricate an improvised nuclear or radiological explosive device, has worried people. The director believes that, 'Public concerns are often misconceived and exaggerated. There have been no terrorist attacks against any nuclear facilities in the United Kingdom

and no known examples of malicious activity that could have caused a nuclear explosion or a serious release of radioactivity.' Nevertheless, the director concedes in his 2002 report that 'a successful sabotage attack on a nuclear facility could cause widespread radioactive contamination and loss of life'. He adds that the theft of nuclear material could also have serious consequences and that, 'It is essential, therefore, that stringent security precautions are taken by the civil nuclear industry, well above normal commercial standards.'

The Nuclear Industries Directorate of the United Kingdom's Department of Trade and Industry says that the department participates in the international debate about standards of physical protection and attempts to ensure through its regulations that UK civil nuclear material and facilities meet international regulations and guidance on security measures. The department is also represented on the Police Authority of the United Kingdom Atomic Energy Authority Constabulary, which patrols UKAEA and BNFL sites.

Site Plans and Vetting
United Kingdom civil nuclear operators must have site security plans and transport plans in order to meet the statutory requirements for the protection of nuclear materials and facilities. These requirements range, for example, from physical protection features such as fencing, CCTV and turnstile access, to the roles of security guards or the United Kingdom Atomic Energy Authority Constabulary at the more sensitive sites, to the protection of proliferation sensitive data and technologies, and the trustworthiness of the individuals with access to them.

The director of the OCNS assumes that any attacks by proliferating states or terrorist groups will be planned carefully in advance. He has written in his 2002 report that,

> Although most individuals working within the industry are reliable and trustworthy, those planning attacks may seek to use a disaffected or suborned

insider with exploitable access. Attempts by criminals to obtain saleable material or information must also be circumvented. In addition, individuals may pursue harmful or irresponsible activities on their own account, perhaps in ignorance of the possible consequences.

For these reasons, OCNS supervises a comprehensive system of security vetting applicable throughout the industry. The system must accommodate human rights and data protection legislation. The challenge involved in vetting personnel is considerable. For example, in the twelve months to March 2001, OCNS issued 9,178 vetting clearances and 3,356 revalidations. The director of the OCNS describes the pressure of this task as 'unremitting' and the sheer scale of what is involved suggests that it might well be possible for a determined individual to circumvent this intended barrier to those intent on mischief. All visitors to the Sellafield plant must provide basic personal details and passport numbers in advance of their visits so that they, too, may be vetted by security forces and issued with appropriate passes.

Stealing Plutonium

But, what is there to stop an individual or group from stealing plutonium that is in transit between plants, and using it to make a nuclear bomb? There are strict safeguards in the United Kingdom relating to the transportation and uses of plutonium. BNFL can only export plutonium that comes from reprocessing if it has permission from the government of the United Kingdom. There are also international rules and agreements regulating how exportation is conducted. BNFL claims that at all stages, from leaving the reactor, through the reprocessing stage, to the making of new fuel before it goes back into the reactor, the location of plutonium is carefully monitored. The company also points out, reassuringly, that,

> The plutonium used to make nuclear weapons has a high percentage of the Pu-239 isotope, which is

created [only] when reactors are run in a particular way just for the purpose of making plutonium for weapons. Plutonium that is produced when reactors are operated for civil electricity generation has a very different mixture of plutonium isotopes.

BNFL notes that,

It has been claimed that if the plutonium taken from used fuel fell into the wrong hands, it could be used to make a nuclear weapon. Plutonium that is taken out of used fuel, from civil nuclear reactors by reprocessing, is kept under strict safeguards. Nuclear weapons are not made from safeguarded plutonium.

Nevertheless, BNFL concedes that,

it may still be technically possible to use the plutonium from civil reactors to manufacture an explosive device. However, anyone wanting to make this type of device would need access to safe handling facilities, a great deal of skill and specialist knowledge, and access to very special technology and parts which are not generally available.

It says that,

Even if all these things were available, it would take months, or even years, to make an explosive device from civil reactor plutonium, and it would be extremely difficult to do in secret.

Assurances about the safety of plutonium and the security that surrounds it, are thus somewhat qualified. This is so even in a democratic state such as the United Kingdom, which has many financial resources and a normally reliable security system.

With hundreds of reactors operating in more than thirty countries around the world, the scope for plutonium being stolen and falling into the wrong hands is increasing all the time. Moreover, the possibility of a terrorist organisation acquiring plutonium and combining it with conventional explosives to make a 'dirty bomb' are also feared. The consequences of having plutonium spread in such a manner are difficult to assess. Sellafield will remain a potential source of plutonium for terrorists and thieves, notwithstanding the high level of security that exists at the site.

Good Morning, Constable

The United Kingdom Atomic Energy Authority Constabulary (UKAEAC) was set up as part of the UKAEA by the Atomic Energy Authority Act 1954. It was created to protect nuclear materials on UKAEA sites and in transit. Later legislation extended its remit to BNFL and Urenco sites when these bodies were created. The constabulary may deploy armed police officers to protect nuclear material. It provides perimeter and internal patrols, including dog patrols, and a police presence at all open perimeter gates and at certain designated sensitive buildings and areas inside sites. 'Unarmed UKAEAC police officers guarding the periphery of civil nuclear sites are generally able to deal with any minor public order incidents involving anti-nuclear demonstrators without having to wait for assistance from county forces, or risk involvement by armed officers', says the director of the OCNS in his 2002 report. The constabulary is also authorised to investigate criminal activities inside sites, and the evidence that it discovers can be used to revoke an individual's security clearance.

One aspect of the constabulary that attracted some critical attention from its foundation was that a special Police Authority, representing the prime users of its services, had the power to agree on the force's funding and decide its policing plans and objectives. In 1998, the House of Commons Trade and Industry Committee concluded that there had been pressure to reduce police numbers at several sites to save costs. Left to regulate its own policing, the nuclear industry had cut corners to save money, thereby increasing the risk to the public from any breach of security. However, a degree of independent regulation of security in the civil nuclear industry was subsequently provided by the Directorate of the Office for Civil Nuclear Security. The present OCNS director set up a Standing Committee on Police Establishments (SCOPE), shortly after taking up his job and in the wake of a public dispute, in 1998, about police numbers at the Dounreay site. He later wrote in his report of 2002 that the 'SCOPE' committee was 'created to reassure ministers, parliament and the general public that policing levels at sites protected by the UKAEAC would be determined in future by my office on security grounds and not by the companies' commercial interests'. He added that the number of constables had been increased by 27 per cent against the baseline 1998–99 establishment (478), excluding a supernumerary complement of 42 police officers to escort MOX shipments.

The importance of being able to rely on a police force that puts security before profits has been underlined by a further decision of the United Kingdom Secretary of State for Trade and Industry, on 28 November 2001. He decided to detach the constabulary altogether from the UKAEA and to re-constitute it as a stand-alone force independent of the nuclear industry and under a new, statutory Police Authority, subject to the necessary legislation. The Anti-Terrorism, Crime and Security Act 2001 extended the jurisdiction of the UKAEAC, strengthened sanctions against the disclosure of sensitive nuclear information and technology and brought in powers for reform of the civil nuclear security regime through secondary legislation.

There are about 500 officers in the constabulary, and it is accountable to parliament through the Department of Trade and Industry. Further information on the force is available on its website at http://www.ukaea.org.uk/ukaeac/index.htm.

Security Inspections
No matter how many threat assessments are made by the OCNS, no matter how high the standards are set and how

well the personnel are vetted, there will always be a need for the OCNS to visit sites and to see for itself how well guarded they are. The director has a handful of inspectors to do this job, which appears to be a big improvement on the position prior to 1996. Before then, states the director in his 2002 report, 'the inspectorate had been reduced to the point where it could not sustain a regular programme'. Nevertheless, given heavy demands since 11 September 2001, the director felt obliged to complain, in March 2002, of 'staff shortages and other pressures' and admitted that the monitoring of earlier improvement requirements at nuclear sites had been 'temporarily' interrupted. The OCNS security inspectorate ought not to be confused with the Nuclear Installations Inspectorate of the Department of Trade and Industry, which deals with health and safety matters. Since 1996, it is said, OCNS has completed inspections at every British nuclear site subject to regulation, covering all aspects including information and personnel security procedures, as well as physical security and guarding. According to its director, 'A comprehensive series of inspections has also been concluded covering all areas at the large Sellafield complex and its satellite [dump] at Drigg. In addition, a wide range of security improvements at sites has been initiated, most of which have now been completed.'

The Security Report of June 2002

How exactly Sellafield and other nuclear sites are protected against possible terrorist attacks is a matter of great secrecy. Nevertheless, some insight into the security arrangements for Britain's nuclear facilities is afforded by the important and wide-ranging official report that was published by the director of the OCNS during June 2002, to which a number of references have already been made. Presented to the United Kingdom's Secretary of State for Trade and Industry, the report provided information on the state of security in the civil nuclear industry and the effectiveness of security regulation.

The report is made all the more interesting by the fact that the director includes some background information about nuclear security regulation that had not been made available before. It also addresses issues raised by the attacks in New York and Washington, which, as the director acknowledges, 'provoked widespread public concern about the vulnerability of nuclear facilities to terrorist attack'. The considerable difficulties of making nuclear sites secure are acknowledged. As the director sees it,

> Those contemplating attacks may have the advantage of surprise, selecting the time, method and location best suited to their purposes. Nuclear facilities are large industrial complexes which cannot operate without substantial numbers of workers and regular deliveries of supplies, complicating effective controls on access. The amount and sensitivity of nuclear material varies between sites, new construction and maintenance work sometimes involving numbers of contractors have to be catered for, and some sites need to attract tenant organisations. Furthermore, for business reasons and to meet various statutory requirements, a considerable amount of information about these sites and about nuclear technology is made publicly available.

The director points out in his report that security is about more than physical barriers and guards. Computers and safety-critical electronic systems inside sites have to be protected against hacking and other forms of interference.

> In addition, nuclear material has to be transported for a variety of reasons, usually within the country by road or rail, and abroad by sea or, in certain limited circumstances, by air. It helps that most sites are located within tight-knit local communities: workers and their families and the local police are encouraged to report strangers behaving suspiciously.

According to the director, sites are protected by a range of interlocking physical barriers, technical systems and security procedures. Frequent and random security patrols, both inside and outside sites, on foot, in vehicles, and with trained police dogs, are a valuable deterrent. The most sensitive sites are protected by armed police officers. Staff, visitors and vehicles are subject to search or may be scanned electronically. Sensitive areas and systems within sites are given additional protection, providing defence in depth. These measures in combination are designed to deter or counter any form of ground-based attack, assessed as falling within the capabilities of terrorist groups and others posing a threat. They are also designed to reassure workers inside sites that they are not at risk. Other measures are taken to safeguard nuclear material in transit and to protect civil nuclear sites from attack from the air.

The director observes that the challenge of keeping sites secure is made more difficult by the 'irresponsible' behaviour of certain kinds of protestors. Could he mean bloody Greenpeace? Bloody Irish environmentalists?

Designer Threats
The United Kingdom has recently developed a new methodology for assessing possible dangers to the security of its nuclear sites. According to the director of the OCNS, over the eighteen months to June 2002 (the period that included the terrorist attacks of 11 September), his office has evolved a fresh procedure to assess security threats, incorporated in a key planning document known as the 'Design Basis Threat (DBT)'. The document is based on intelligence about the motives, intentions and capabilities of potential adversaries. It is designed to provide a definitive statement of the possible scale and methods of attack that could be faced at civil nuclear sites, or when nuclear material is being transported. This 'DBT' excludes possible security threats and methods of attack that are judged not to be relevant to the civil nuclear industry in the United Kingdom. The DBT also takes account of the

availability of countermeasures and contingency arrangements provided by the police, the Ministry of Defence and other agencies. 'For obvious reasons', observes the director in his 2002 report, 'the assessment is classified SECRET and no further details can be published.' The DBT document now provides the basis for the design, implementation and management of security measures and systems by the regulated civil nuclear companies. It is also being used by OCNS to develop or revise mandatory and discretionary security standards and guidance, to evaluate site and transport security plans prepared by the operators, and to monitor compliance. This criteria-based approach is regarded by the International Atomic Energy Agency as best practice and is also being adopted by regulatory authorities in other countries.

All of which seems to be a way of saying that the director is sure that the British authorities are doing their best to keep us safe, but that the details of how they are doing so are top secret. Critics note that the DBT document has also been written to reduce pressure on the OCNS and to allow BNFL and other companies a greater degree of discretion in the operation of their plants. They fear that, while official control over the constabulary has been asserted, some forms of operational regulation may be weakened. However, the director of the OCNS says that his aim is 'to encourage the companies to undertake their own site audit programmes, to suitable quality standards, the results of which I would be prepared to take into account in my office's inspection process'. He maintains that,

> This will free OCNS Inspectors to drill down in greater detail on particular issues and undertake spot checks. In addition, in a move away from the present prescriptive regime of standards and regulations, the companies are being allowed to propose alternative arrangements, providing comparable levels of security but more closely reflecting company policy or local circumstances. The Design Basis Threat document has been written partly with this aim in

mind, to provide the necessary foundation. The combination of this extra investment in effort and more flexible, focused working arrangements should benefit all the stakeholders involved, enhancing both security at sites and regulatory oversight.

One certainly hopes that the director is right.

International Co-operation

The United Kingdom is a party to the Convention on the Physical Protection of Nuclear Material, which came into force in 1987 under the auspices of the United Nation's International Atomic Energy Agency (IAEA). The Convention primarily obliges those countries who have signed it to meet defined standards of physical protection for international transport of nuclear material and to co-operate in the recovery and protection of stolen nuclear material, and it also promotes international co-operation in the exchange of physical protection information. Responsibility for establishing and operating a comprehensive physical protection system for nuclear materials and facilities in domestic use, storage or transport rests with the government of each state. However, the IAEA's document entitled 'The Physical Protection of Nuclear Material and Nuclear Facilities' provides a useful basis for guiding states and is available on the IAEA website. The United Kingdom claims to follow this guidance closely in its own domestic regime.

The United Kingdom has also assisted the IAEA for some years in a number of programmes designed to improve standards of security worldwide. Experts from the OCNS have been provided for the 'International Physical Protection and Advice' missions, and for training courses on physical protection run by the IAEA. The United Kingdom, through OCNS, has also provided some bilateral assistance to a few countries to upgrade the physical protection systems at a number of their more sensitive facilities.

On 1 August 2002, the Energy Minister of the United

Kingdom, Brian Wilson, announced the awarding of
contracts to BNFL, PE-International Consulting Ltd and
RWE NUKEM Ltd, to manage United Kingdom assistance
for the first in a portfolio of complex projects to help address
the nuclear legacy in the former Soviet Union (FSU). Wilson
said that, 'The nuclear legacy of the FSU is one of the most
important challenges facing the international community.
The considerable environmental, security and proliferation
threats it presents do not respect international boundaries
and pose a direct threat to us all.' Wilson recalled that at the
exclusive G8 summit in Kananaskis, Canada, in June 2002,
Prime Minister Tony Blair had committed up to $750 million
over ten years to support the G8 'Global Partnership Against
the Spread of Weapons and Materials of Mass Destruction'.
The focal point for information on UK assistance to
managing the nuclear legacy in the FSU is a new website:
http://www.dti.gov.uk/energy/nuclear/safety/fsu.shtml. The
European Commission has also proposed a range of measures
relating to the nuclear legacies of the former Soviet block.

These British and European initiatives provide new
opportunities for BNFL. On 25 November 2002, BNFL
announced that it had opened an office in Moscow, 'to
continue to build links with Russian officials and nuclear
organisations'. Speaking on that occasion, Charlie Pryor, chief
executive of BNFL's Nuclear Utilities Business Group, said,

> We share a common expectation about the future of
> nuclear energy and a firm belief that nuclear power is
> the only way to generate secure, reliable and
> environmentally friendly electricity. We also share an
> understanding of the value of nuclear fuel recycling
> and a commitment to safety and non-proliferation.

In this and earlier chapters, the nuclear industry's general
'commitment to safety' has been considered. It is time, now,
to turn to what Pryor called 'the value of nuclear fuel
recycling' and, in particular, to the function and safety of
BNFL's reprocessing plant at Sellafield.

CHAPTER 10

Reprocessing

Sellafield is principally a reprocessing and fuel manufacturing plant, and not a station for generating electricity. The most recent manifestation of its reprocessing prowess is the giant THORP plant, which will be described and discussed in detail in the following chapter. The accumulated and dangerous radioactive wastes stored at Sellafield arise largely from its reprocessing activities for British and other companies, as well as from earlier work for the civilian and military atomic programmes.

New Fuel from Old
BNFL offers its British and overseas customers an opportunity to send their used fuel to Sellafield to be made into yet more fuel. The storing and recycling of this spent fuel is a major part of the company's business on its Cumbrian site.

The nuclear fuel in a working nuclear reactor becomes less efficient after a number of years and must be replaced with fresh fuel. The average life of a fuel rod is about four years. Because of its unique characteristics, it has to be taken out of the reactor even though the vast majority of it has not been burnt. Its owners can now dispose of it into storage, for an indefinite number of years, or try to retrieve the unburned portions of the fuel by means of reprocessing. Thus, the major components of the used fuel (namely uranium and plutonium) can be reused after they are separated from the

The perimeter fence of Sellafield.

One of the approach roads to Sellafield. © KARE

useless waste components (i.e. from certain 'fission products'). This separation is achieved by means of 'reprocessing'. Reprocessing involves removing the metal casing from around the fuel and dissolving the fuel itself in hot, concentrated nitric acid. The uranium, plutonium and waste which are dissolved in this way are then separated from each other using a number of chemical processes.

BNFL notes today that, 'Used nuclear fuel still contains large amounts of energy. Once we have removed the used fuel from the reactor, we can either dispose of it as nuclear waste or reprocess it to separate the 96 per cent of uranium and 1 per cent of plutonium from the 3 per cent waste.' The uranium is depleted and needs to be re-enriched before being used again. That 1 per cent of plutonium is particularly valuable. By mixing it with new uranium from Springfields, a fuel is produced at Sellafield that is more efficient than enriched uranium on its own. This new product is mixed oxide fuel (MOX).

Same As It Ever Was

In the early 1950s, reprocessing began at Sellafield in order to extract plutonium from spent fuel for the purposes of building and exploding an atomic device. In the early 1960s, a second reprocessing plant (B205) came on stream in order to reprocess fuel from the first civil nuclear reactors, the Magnox reactors.

In 1983, having investigated an incident of radioactive contamination on the beach at Sellafield, the United Kingdom's Health and Safety Executive noted that, 'The principal function of the Sellafield site is the reprocessing of irradiated uranium fuel for the recovery of uranium, the separation of plutonium, and the storage of the highly active fission products which form the main radioactive waste material in the irradiated fuel.' By the early 1990s, work was progressing on plans for THORP, the third reprocessing plant to be built on the Sellafield site. THORP was planned to process oxide fuels from the AGR and PWR reactors that succeeded the earlier Magnox reactor.

In 1996, Dr Bill Wilkinson, the former Deputy Chief Executive at BNFL from 1986 to 1992, contributed to a British television documentary entitled 'Inside Sellafield'. He said that, 'There are really two options [for disposing of spent nuclear fuel]. One is to reprocess and recycle, and the other is to directly dispose of spent fuel. The UK concept has always been reprocessing and recycling, and that, really, is what Sellafield is all about. It's a reprocessing plant.' In 2001, the authors of the 'WISE' report for the European Parliament noted that, 'Between 1965 and the end of year 2000, about 26,000 tonnes of spent gas graphite fuel were reprocessed by the B205 line at Sellafield. About 3,000 tonnes of spent light water reactor fuel have been reprocessed at THORP since 1994.'

While the central reprocessing facility at Sellafield today is the new THORP installation, fuel from Britain's Magnox reactors continues to be reprocessed there at the facility designated as 'B205'.

Risk and Danger

In 1978, the Parker Inquiry concluded that reprocessing at Sellafield was in the interests of the United Kingdom because it decreases dependence on foreign energy resources and on fossil fuels. It also reduces, wrote Parker, 'the risk of the escape of more plutonium than is necessary' and allows for the remaining spent fuel to be stored more safely as solid glass (i.e. in vitrified form) than as a liquid. His reference to a 'necessary' escape of plutonium was frankly utilitarian.

There are many dangers involved in handling radioactive waste, and these dangers are never less present than when spent fuel is being reprocessed. When spent fuel arrives at Sellafield for reprocessing, it is first stored underwater. Special measures are taken to protect the workforce at the fuel-handling plant and elsewhere on site. Operations are carried out remotely behind shielded walls, and there are extensive ventilation and decontamination facilities.

In 1990, an official United Kingdom White Paper on the environment confirmed that reprocessing allows over 95 per

The events of 11 September 2001 increased fears of a terrorist attack on Sellafield. (Allsport UK)

Sellafield, with the steaming stacks of Calder Hall power station. These stacks long powered Sellafield, but are now being shut down and decommissioned, and energy for the site will come from further afield.
© *Greenpeace/Alan Greig*

THORP: The mighty Thermal Oxide Reprocessing Plant is at the heart of Sellafield and at the centre of controversy surrounding Britain's nuclear fuel industry. © Greenpeace/Steve Morgan

Jack Allen (centre), Head of the MOX Operations Unit at Sellafield, takes the media on a tour of Sellafield's MOX plant, 2002.

Radioactive fuel is loaded into big flasks at Sellafield, and is transported by train to the nearby port of Barrow-in-Furness.

Dermot Ahern, Ireland's Minister for Communications, Marine and Natural Resources, warned in 2002 that the Irish Sea is 'a nuclear fuel highway – the final destination for other nations' nuclear fuel'.

Captain Malcolm L. Miller, BNFL's Head of Operations Transport, being interviewed on the dockside at Barrow-in-Furness. Behind him is the Pacific Crane, a nuclear carrier used by BNFL.

September 2002: The Rainbow Warrior *departs Dublin for the Irish Sea to protest against the ship* Pacific Pintail *carrying nuclear waste to Barrow-in-Furness, for delivery to BNFL Sellafield.*

John Swinney, MSP, leader of the Scottish National Party, which 'fully supports the Irish government in their attempts to stop the UK government polluting the Irish Sea with nuclear waste'. (SNP)

Martyn Turner has commented wryly on the nuclear debate for a quarter of a century. In the early 1970s (above), Ireland itself planned to build a nuclear power plant at Carnsore in Co. Wexford. © Martyn Turner

More recently (below), Ireland's own environmental practices have left something to be desired. © Martyn Turner

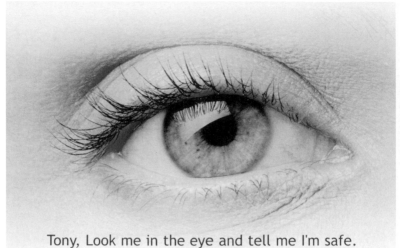

Tony, Look me in the eye and tell me I'm safe.

In April 2002, more than one million postcards (above and below) were mailed from Ireland to England as part of a 'shut Sellafield' campaign. John Hume MP told the House of Commons that it was the 'largest amount of correspondence from individuals on one subject that any Prime Minister has ever received'. The cards were also addressed to Prince Charles and to Norman Askew, who is due to step down as chief executive of BNFL in the summer of 2003.

Tell us the truth.

cent of the uranium and plutonium present in the spent fuel to be reprocessed. However, some highly radioactive waste is also created in liquid form and must be stored. This waste is greater in volume than the spent fuel itself prior to reprocessing, and it presents a storage problem. Yet, reprocessing itself is 'a proven technology', according to the government of the United Kingdom.

During public consultations in the 1990s, the United Kingdom government pointed out that reprocessing has been going on in some form at Sellafield for many years. It claimed that, 'Reprocessing does not create additional radioactivity. Indeed, it removes the radioactive plutonium and uranium, which would otherwise form part of the waste, and allows these materials to be recycled.' The United Kingdom government stated that such recycling can reduce waste volumes in other areas of nuclear production by eliminating the need for additional mining, milling and enrichment of new uranium to replace that which is spent. It states that, 'Reprocessing is the only long-term spent fuel management and conditioning strategy that is being actively pursued at present in the United Kingdom.' France, too, has civilian reprocessing facilities. Japan is building some.

'Efficient' and 'Safe'

BNFL has listed 'several good reasons for reprocessing nuclear fuel'. The company argues that it is 'efficient' and that,

> For nearly forty years reprocessing has been shown to be safe and to have only a very minor effect on the environment. For example, at our Sellafield site, where reprocessing is the main business, the dose of radiation that a person living near Sellafield would receive in a year would be about the same dose as you would receive on a return flight to the Far East. This is because when you are flying, you are exposed to more naturally occurring radiation.

BNFL has a large stockpile of used fuel. The company says that, 'If this used fuel is recycled into new fuel, it will generate enough electricity to meet all of the UK's electricity needs for about two years.' Already, down the years, 15,000 tonnes of uranium extracted from used fuel from the United Kingdom's old Magnox power stations has been recycled and made into new fuel for Britain's Advanced Gas-Cooled Reactors. In all, to date, BNFL has reprocessed over 40,000 tonnes of fuel for UK and overseas customers.

The reprocessing of spent fuel means that it is not necessary to mine as much fresh uranium as otherwise. However, BNFL states that reprocessing allows the world's uranium supply to last longer, which suggests that it envisages an equal amount of uranium being mined in the long run. Their point is that, 'To argue that there is already plenty of uranium in the world and no need to recycle it is like arguing that there is plenty of sand in the world, so we should stop recycling glass which is made from sand.' According to BNFL, if used fuel is not reprocessed, 'the only other short-term option available is to keep it in stores above ground'. Eventually, after storing the used fuel, a decision will still have to be made either to reprocess it or to bury it deep underground.

BNFL has long been required to put a term into contracts which obliges customers to take back waste products which result from the reprocessing of their spent fuel. Such terms have been inserted in contracts since the 1970s. However, in practice, the waste has not been returned. Instead, these highly active and dangerous substances have been kept at Sellafield, adding to the volume of radioactive waste that is already stored there. The delay has been caused partly by the failure to proceed more quickly with plants capable of turning liquids into a hard glass or ceramic form – the process known as 'vitrification'. A few shipments of vitrified waste have been returned from the French reprocessing plant at Cap de la Hague to customers abroad.

During 2001, when making observations on BNFL's continuing release of radioactive technetium-99 from Sellafield into the Irish Sea, the Irish government recalled that, 'The OSPAR Decision 2000/1, adopted at the meeting of the OSPAR Commission in June 2000, recognized that nuclear reprocessing facilities are the dominant sources of radioactive discharges, and that the implementation of the non-reprocessing option for spent fuel would produce substantial reductions of such discharges.' This decision was a clear signal to those involved in nuclear reprocessing of the concerns felt by the other Contracting Parties about the environmental and public health impact of such activities and related discharges.

Dry or Wet?
Debates have long raged between those who support the reprocessing of spent fuel and those who believe that it is preferable to store spent fuel. While reprocessing allows some fuel to be reused, it is a wet process, during which the old fuel is given a chemical bath. Much additional waste is thus produced in liquid form, some of which is eventually discharged into the sea and some of which is so highly dangerous that it must be stored indefinitely in tanks until it can be vitrified. By not reprocessing, it is possible to keep spent fuel dry and to store it dry.

BNFL argues that, 'By reducing the amount of fresh uranium that is mined, we are lowering the radiation dose which the world population receives and we are producing less waste than if the used fuel was simply disposed of and replaced with new uranium.' However, the authors of the 'WISE' report for the European Parliament concluded that,

> non-reprocessing options, and available dry storage technologies in particular, are considerably less expensive than reprocessing. In addition, their social and political acceptability are much greater than reprocessing. Nuclear utilities are increasingly moving towards dry storage solutions. . . Direct disposal options also significantly reduce waste volumes to be

disposed, due to the large volumes generated by reprocessing.

The government of the United States stopped reprocessing fuel from civilian reactors as long ago as the 1970s, but the governments of France, Japan, Russia and the United Kingdom still consider it an acceptable option. German plans to build a reprocessing plant at Gorleben in Lower Saxony were abandoned due to public opposition.

Not Convinced
During 2000, when making observations on the proposed 'United Kingdom Strategy for Radioactive Discharges 2001-2020', the Nuclear Safety Division of the Irish Department of Transport made clear its opinion that, 'it is essential that the justification of nuclear reprocessing (which is the dominant source of marine radioactive discharges) be reviewed in the light of current circumstances'. The department argued that current British targets for the reduction of annual 'beta/gamma' discharges from Sellafield were not compatible with the international OSPAR Strategy, which will be considered later and which aims to achieve 'close to zero' concentrations in the marine environment.

In October 2001, when the government of Ireland repeated its opposition to the Sellafield MOX plant, it noted that the MOX plant will continue to be dependant for its supply of materials on the THORP reprocessing plant. The Irish government stated that it is 'strongly opposed to nuclear reprocessing activities which generate gaseous and liquid radioactive discharges which contaminate the terrestrial and marine environment'.

Also, during 2001, the authors of the controversial 'WISE' report for the European Parliament noted that, notwithstanding the alleged benefits of reprocessing,

> only 5 per cent to 10 per cent of world annual spent fuel arisings is submitted for reprocessing, with the rest stored pending final disposal in a repository. The

largest centres in the world for commercial reprocessing remain Sellafield in the UK and La Hague in France. Reprocessing involves the dissolution of the spent fuel in boiling concentrated nitric acid and subsequent physico-chemical separations of uranium and plutonium.

They continued by claiming alarmingly that,

Multiple waste streams are created by these physical and chemical processes. While some wastes are retained and conditioned, considerable volumes of liquid and gaseous wastes are released to the environment. Reprocessing operations release considerably larger volumes of radioactivity than other nuclear activities, typically by factors of several 1,000 compared with nuclear reactors.

The 'WISE' authors maintained, among their other findings, that

the reprocessing of spent nuclear fuel at Sellafield and La Hague leads to the largest man-made releases of radioactivity into the environment worldwide. The releases correspond to a large-scale nuclear accident every year. Some of the radionuclides released in great quantities have half-lives of millions of years.

They stated that, in the United Kingdom, about 90 per cent of nuclide emissions and discharges from the UK nuclear programme result from reprocessing activities. They alleged that the past discharge of certain key radionuclides, at Sellafield and La Hague, and further planned releases, constitute a violation of the letter and spirit of the OSPAR Convention. The authors claimed that the European Commission does not effectively use its verification rights, but

is highly dependant on information provided by Member States and is therefore apparently not in a position to guarantee that the Basic Safety Standards are respected concerning the La Hague and Sellafield facilities.

The 'WISE' report concluded that,

There are great uncertainties involved in the assessment of doses to populations and subsequent health effects. The release of large quantities of long lived radionuclides at Sellafield and La Hague therefore violates the Precautionary Principle, laid down, *inter alia* [among other things], in the European legislation, Agenda 21 and the Earth Charter of March 2000.

As already noted, the British government and BNFL dismiss the 'WISE' report as unreliable.

CHAPTER 11

Touring THORP

T he reprocessing of old or 'spent' nuclear fuels, which is the core business carried on at Sellafield, is clearly a dangerous and controversial activity. The Thermal Oxide Reprocessing Plant (THORP) is a massive structure that represents the third generation of reprocessing by BNFL. It has cost the United Kingdom about £3 billion sterling to build THORP and its associated facilities. As noted above, earlier reprocessing plants have served both military purposes and the needs of the civil nuclear reactor programme.

With up to 8,000 workers employed on its construction, THORP is said to have been, for a while, the biggest building project in Europe. THORP is known as an 'island' facility because of its self-sufficient nature. THORP was built to reprocess oxide fuels from certain Advanced Gas-Cooled Reactors and from Pressurised Water Reactors, mainly located overseas. Its construction was permitted following a public enquiry. It is earning significant foreign revenue for the United Kingdom.

Visiting THORP
THORP is a striking structure. Visiting it, in 2002, felt a bit like stepping onto the set of *Star Trek*. But there was one very real difference between my visit to THORP and a tour of some Hollywood studio. Plutonium! I could not help wondering if, just possibly and despite sophisticated safety

checks, a little speck of plutonium might not have found its way onto the floor of THORP and, off the floor, into my skin or lungs.

THORP is an enormous, modern, box-like structure, which is nicely painted on the outside in beige and brown, with highlights picked out in red. Almost one-third of a mile long, its corridors connect an amazing array of huge halls and deep open pools in which are stored great bottles of radioactive spent fuel.

But, be prepared! You need blue woolly socks to get into THORP. Never mind the advance security clearance, the special badges, or the clip-on radiation monitor. You know that you have really arrived at THORP when they hand you a pair of woolly socks in the BNFL company colours. You are asked to remove your shoes and don the regulation woollies.

You must then place your shoes, with any mobile phone and other items of your choice, in special lockers. Your locker has a key, which seems surprising given that you have already passed a security screen, and bearing in mind that all employees of BNFL are vetted by the security forces. Might there be persons in here who would steal your shoes, but who can be depended upon not to nick a single gram of plutonium?

So, you put the blue socks on over your own socks, or over your bare feet if you were not already wearing socks. This is mainly for reasons of 'hygiene', we are told. Then, in stockinged feet, you proceed to an adjacent room to pull on the next layer of futuristic, protective clothing. This is a cotton, knee-length coat, with a BNFL logo. Again, the garment is blue; but it has a yellow collar. It looks like it was manufactured in the same place that has been making coats for assistants in hardware stores since a very long time ago.

There is no white, glossy space-suit or mask to be worn, which is a bit disappointing after everything one has read about the dangers of nuclear power. Apparently, only those who must go into the inner and most dangerous recesses of the THORP sanctum wear such glamorous items. If BNFL really wants to impress the public, then the company should camp up its clothing. BNFL might earn a few pounds sterling by following the example of Disneyland and offering to sell its visitors on their way out a photograph, taken automatically in a dramatic-looking location, showing them in fancier protective garb than is currently worn. But, then again, that might frighten the folks back home.

It is deemed imperative that the blue coat that is issued to visitors be 'buttoned up' completely, although it is a most modest kind of thigh-length garment. The top patent fastener on my coat did not work properly and kept opening, but I appear to have survived without receiving a deadly dose of radiation.

Then, after the coat is secured, it is forward to the shoe counter. This counter is just the right height to sit on comfortably and to swing your legs over. Here, they place white shoes in your size (sorry, no halves) across on the other side. You must swing and place your feet down into these shoes without letting the blue socks touch the ground. Otherwise, it will be back to square one to get a new pair of socks. The shoes resemble those in bowling alleys, where (come to think of it) the provision of clean socks to customers for hygiene purposes would be a real scientific advance, for which idea the world might be eternally grateful to BNFL.

Any equipment that you have been permitted to take into THORP must now be placed on the counter where it is inspected and monitored by staff. And then it is onwards and upwards into THORP, provided that you swipe your BNFL security card through the special control gates that are encountered every so often.

Preparation Area
The first hall that one enters is a bit of a disappointment. Admittedly, the high ceiling of this great Preparation Area is quite impressive, with its great, yellow beams. However, nothing much seems to be happening. I soon realise that, in fact, nothing very visible *is* happening. But there is plenty

going on within the radioactive fuel that is invisible. THORP is a graveyard, where large quantities of spent fuel are entombed for a certain period while it decays, before BNFL breaks down what remains into uranium, plutonium and useless waste.

On one side of this Preparation Hall, a big, corrugated transport flask sits upright on its end. It had earlier been brought to Sellafield on its side, by rail. Standing the flask upright allows the shiny steel cylinder of used fuel, inside it, to be removed. This cylinder is also known as a 'multi-element bottle'. The flask and cylinder weigh up to 110 tonnes, while the spent fuel within them is said to weigh just four or five tonnes. On the other side of the Preparation Hall, a group of such cylinders huddle, like silent prisoners confined within a steel cage.

When the two members of the television crew in our group prepare to film the Preparation Area, it is clear that they will not be given a free hand. BNFL's Media Affairs Manager must first vet their planned shot. The same criterion applies when I wish to take a still photograph. This restriction appears to be as much for reasons of commercial confidentiality as for security. We shall be reminded a number of times during our visit to the THORP and MOX plants that nuclear power is a very competitive business.

Much of THORP seems strikingly empty of human life. Such functions as are required are highly automated, and the number of people needed to carry out operations is said to be 'minimal'. We are impressed to learn that two Euratom inspectors are visiting this morning, who will inspect any and all movements on site. Euratom is an international organisation of the great and the good involved in nuclear power and it surely has very high standards, has it not?

We are less impressed to learn that the Euratom inspectors normally come on certain days, and that BNFL knows when they are coming. For some reason, I am struck by an image of farmers rejoicing upon hearing that all future inspections by the Department of Agriculture will be at scheduled times.

However, it is said that Euratom also has cameras and other detection equipment installed permanently at THORP, and that BNFL has no access to that equipment. Now I think of the Department of Agriculture using satellite images of farms to check applications for grants, which it does, or installing CCTV in the milking parlour, which it does not.

Inside THORP, the cylinders of spent fuel must be stored once they have been removed from their flasks in the Preparation Hall. Their storing under water lasts for at least five to six years in order to allow their radioactivity to dissipate to a level where it is commercially viable to recycle the used fuel, which is 96 per cent depleted uranium, 1 per cent plutonium and 3 per cent useless waste. At first, the cylinders that have been removed from their flasks are stored briefly in a deep pool in the Preparation Area itself. Every day, through this pool, one thousand cubic metres of water are circulated. The cylinders are later moved from this first pond into the main storage pond, nearby, which is an awesome sight.

Storage Pond

Climbing stairs, we enter the storage pond hall from on high, at a point about halfway between its floor and roof. There, we stand on a metallic platform from which we gaze down upon a stretch of water that is the size of two average football pitches. This is the centrepiece of THORP, the still point of the slowly turning world of reprocessing at Sellafield.

Filled with water to a depth of eight metres, the storage pond contains row after row of steel cylinders, each cylinder full of radioactive waste. Because of the accumulation of waste at Sellafield, some of these cylinders are up to 20 years old, having been kept closed for far longer than the five or six years normally needed to allow their radioactivity to dissipate to a level where it is commercially viable to recycle it. Two thousand tonnes of spent fuel are stored here. The cylinders rise erect under still, dark water in which their steel shines dully. The entire aqueous vault is wrapped in an eerie silence.

At the bottom of the massive storage pond runs a platform about one metre high. It is on this that the cylinders stand, each about four metres tall, although they vary somewhat in size depending on the configuration of the nuclear reactor from which they have been removed. Above the cylinders are another three metres of water. The pond is said to contain a volume of water equivalent to that of twenty Olympic-size swimming pools.

Above the massive storage pond is a great expanse of space, crowned by what looks like a corrugated tin, or stainless steel, roof. There are a few skylights where the roof meets the walls and, through these openings, sunshine streams in. It appears anomalous, like the blue tent of sky at which confined prisoners gaze longingly. As I look up, my mind fills with TV images of jet airliners ploughing into the tallest buildings in Manhattan, images slowed down and repeated on news bulletins time after time.

How well the storage pond roof might respond to a fully laden jet airliner landing on it is a matter for conjecture. BNFL remains tight-lipped about security. Standing there, halfway between two thousand tonnes of nuclear waste and the open air, it felt to this writer a little like being in the loft of a great hay-shed. The roof looked like it would respond probably not very well to such a severe impact. However, BNFL claims that all security eventualities have been carefully reviewed since 11 September 2001, and that the danger of catastrophic damage spreading beyond Sellafield itself is minimal or non-existent.

From the great storage pond, individual flasks are selected after they have been cooled for at least five years. These are then sent forward to the head-end pond, where the spent fuel will be removed for immersion in a chemical bath. The head-end pond is an area where normal operations are more potentially dangerous, because the spent fuel is actively being extracted. It must be remembered that the spent fuel contains plutonium and other 'fissile' or highly radioactive materials.

Head-End Pond

Entering the head-end pond area, you are reminded aurally of the even greater danger that is here, in practice, than in other parts of THORP. Now, and later during our visit to the MOX plant, a sort of sonar sound, consisting of two synchronised signals, is continually emitted at a constant level. We are informed that, in the event of computers detecting any alteration in standard systems of control, this 'Criticality Alarm' signal will be disturbed and prompt an evacuation and other emergency procedures.

The great support beams of the head-end pond are painted a fairly vivid orange and yellow, in glowing contrast to the muted tones of its walls, floor and ceiling. Again, we find ourselves looking down into a deep pond, where various large pieces of machinery lie idle. It is here that Sellafield personnel deploy their remote technology to remove the lids from cylindrical, multi-element bottles of spent fuel, sent to them from the adjacent storage pond. Here, they take the spent fuel out of its cylinder and into a monitoring machine, to make sure that it has gone through the correct cooling period. It is then chopped up and dissolved in nitric acid. Uranium and plutonium within the spent fuel are purified and turned into powder, which can be used to manufacture either MOX or standard uranium fuel.

It is here too that the vast quantities of water pumped throughout THORP end their journey, before being sent on to another plant within Sellafield. The water is said to be very clean and uncontaminated, even though spent fuel is being removed inside machines that operate underwater in the head-end pond. Asked what might happen if someone fell into this pond, a BNFL official replies that one might drown, at worst. Perhaps he is tempted to test his hypothesis on a visitor. A regular lifebuoy, like those on the promenade at Brighton or Bray, is bolted to the handrail above the pool, but nobody takes a dip.

As already mentioned, about 1 per cent of spent fuel consists of plutonium. About 96 per cent consists of uranium. We are told that the plutonium extracted at THORP can go directly to Sellafield's MOX plant. However, the uranium that is extracted is depleted and must be sent to another BNFL facility, to be re-enriched. It may then be returned to Sellafield for making into new fuel if desired. But we are told that, at present, it is fresh uranium from Springfields that is actually used for making MOX fuel at Sellafield. There appears to be a law of diminishing returns which determines that the more that the same uranium is recycled the less good it is.

Once the uranium and plutonium have been extracted from spent fuel at THORP, it is necessary to store or dispose of the residual 3 per cent of the particular lot of spent fuel that has been reprocessed. Chemicals used in reprocessing that spent fuel also remain. This is a most dangerous mix, including Highly Active Liquid which stays at Sellafield pending its intended vitrification and disposal in a long-term facility that has yet to be designed and built at some location yet to be selected. Meanwhile, such wastes constitute a serious hazard at Sellafield, in the opinion of its critics.

From the head-end pond we make our way down into a long, gloomy corridor which leads us towards the MOX plant, which is bolted onto THORP. A description of the MOX plant will be given later.

Justifying THORP
THORP was a long time in the pipeline. It was conceived in the mid-1970s, and the concept was discussed in parliament in 1976. During 1977, BNFL first submitted a planning application for the plant. On foot of that application, an inquiry was conducted by the Hon. Mr Justice Parker for the Secretary of State for the Environment. Parker presented his report on 26 January 1978. There had been evidence from 146 witnesses. Parker was in favour of THORP being built and the government of the United Kingdom agreed with him.

Parker took evidence from scores of witnesses before producing a report favourable to the plans of BNFL. High in his estimation were certain arguments of an economic nature, which supported reprocessed fuel because it would reduce the reliance of Britain on a continuing supply of uranium from abroad, and because it would also reduce reliance on fossil fuels and thus facilitate competition. Critics claim that such economic arguments underestimated the long-term costs of dealing with THORP's consequences for the environment and public health.

Parker also satisfied himself that reprocessing was a safe way of handling spent fuel when compared with methods of storing it. His report notably sets out 'facts' which purportedly demonstrate that plutonium is not as dangerous as people often assume. He was particularly impressed by evidence that plutonium could be sat upon safely with no greater a shield between the object and the sitter than that provided by 'a stout pair of jeans'.

Brian Wynne subsequently published a fascinating critical analysis of Parker's procedures. His analysis is a case study of the manner in which decisions relating to nuclear matters have been taken in Britain. He notes the contrast between previous commitments to THORP and the pretence in public ritual that the issue was free from any such bias. Among other conclusions, Wynne notes that, 'Even political leaders sincerely committed to exploring options could not really escape the belief that, even if postponed, THORP was inevitable. In the end the Minister for Energy reduced the THORP "choice" to a nuclear or a non-nuclear future, which, given the Parker Report's neglect of the latter, was no choice at all [Wynne, *Rationality and Ritual*, 1982, p. 171].'

In 1981, a detailed design for THORP appeared, and construction began a couple of years later. A booklet produced in 1988, for visitors to Sellafield, shows a sketch of THORP dwarfing a side elevation of St Paul's Cathedral, London. The booklet boasted that, 'already the order book for THORP is full for its first ten years of operation, including contracts with many overseas companies'. This was an indirect reference to how the building of THORP had been financed. Its customers invested up front and were credited for a certain volume of orders in return. So great was the Japanese enthusiasm for off-loading its waste on Britain, so significant was Japanese investment in THORP, that the facility was sometimes colloquially known at BNFL as 'the Japan plant'.

By 1992, the main parts of THORP had been completed. In 1993, Peter Melchett, then executive director of Greenpeace and a former Labour government minister at the Department of the Environment, wrote that, 'What will come out of THORP, if it opens, will present far greater disposal problems than what will go into it. The reprocessing of spent nuclear fuel actually mixes the radioactivity into many different chemical and physical forms, spreading the contamination through everything involved in the process – solvents, acids, containers, filters, etc.' He cited a report by a consulting engineer who claimed that the effect of such reprocessing is to increase the waste by as much as 189 times compared with the original spent fuel. Melchett also wrote, in 1993, that BNFL's own figures suggested approximately a ten-fold increase in radioactive discharges into the air and sea in the future. Nevertheless, an official licence to operate THORP was issued that year and fuel began to arrive at Sellafield especially for reprocessing at THORP.

When reprocessing began at THORP, the production of waste at THORP also commenced. From spent fuel they made new fuel, but they also contaminated liquids in the process. In March 1995, active commissioning commenced and spent fuel from overseas was placed in the ponds for the first time. That year, also for the first time, Highly Active Liquid waste (HAL) was sent from THORP to Sellafield facility B215.

Consent for the full operation of THORP was finally granted in August 1997, two decades after its conception. Its overseas customers have included clients in Germany, Sweden, the Netherlands, Switzerland, Spain, Italy, Canada and Japan. The Japanese themselves are constructing a reprocessing plant in Japan, but BNFL expects that the new plant will handle less than one-half of Japanese spent fuel and that the remainder will continue to be sent to Sellafield.

Governments, local authorities, and tens of thousands of individuals have voiced concern at the decision of the United Kingdom government to commission THORP. During 2000, the Nuclear Safety Division of the Irish Department of Transport made its views known on THORP in a written response to a British consultation document on the United Kingdom's strategy for radioactive discharges, dated June 2000. Referring to the requirement in European law that such projects be 'justified' economically and socially, the Irish department wrote that,

> In our view, the document fails to deal adequately with the subject of justification, i.e. the evaluation of the net benefit to society arising from the authorisation of practices resulting in discharges to the marine environment in relation to the health detriment they may cause. No reference to justification is made in Chapter 2 (Principles and Aims) and there is a partial reference to justification in para[graph] 4.2.2 of the document which mentions only all new classes or types of practice. The 1996 Basic Safety Standards Directive (European Council Directive 96/29/Euratom) requires EU Member States to do also the following (Article 6.2): 'Existing classes or types of practice may be reviewed as to justification whenever new and important evidence about their efficacy or consequences is acquired.'

The writer continued,

> In the Department's view there is ample new and important evidence, which would warrant a review of reprocessing activities at Sellafield since 1993 when UK Ministers decided on THORP. This includes

international concern over reprocessing at Sellafield as exemplified by the recent OSPAR decision 2000/1 adopted at the Commission meeting in Copenhagen in June.

In October 2001, the Irish government noted that the controversial manufacture of MOX fuel depended on THORP:

> The MOX fuel to be produced at Sellafield would be made from uranium and plutonium material separated from spent fuel which is reprocessed mainly in BNFL's Thermal Oxide Reprocessing Plant or THORP as it is better known. Therefore, the production of MOX fuel is part of the spent nuclear fuel reprocessing industry at Sellafield.

The extent to which THORP is economically viable is a matter of debate, and is relevant to the British case that it is 'economically justified'. The assessment depends, to some extent, on how one makes the financial calculations. The detail is dense, as may be seen from Mike Sadnicki's 'Examination of BNFL reports and accounts', which he completed in March 2002. This is not a document for the faint-hearted, as it makes few concessions to the casual or lay reader. But it does raise very serious questions about the financial justification case that has been made for THORP and it illustrates clearly the mutual interdependence of THORP and the new Sellafield MOX plant (SMP). As Sadnicki writes,

> The Justification of THORP reprocessing is inextricably bound up with the issue of SMP operation. There is no reason for THORP discharges, unless the SMP uses the separated plutonium. Similarly, there is no reason for the SMP to exist, unless THORP produces separated plutonium. THORP has considerable discharge detriments, SMP less so. But in the current logic, each practice assumes the other as an established fact with

zero cost. In truth, both links in the chain might be losing money. It is clear that the Justification process should be applied to the 'whole system' – to the combination of the two processes, THORP reprocessing and SMP MOX production.

Sadnicki's report was sponsored by Nuala Ahern, Member of the European Parliament, and versions of it have been submitted in the course of the United Kingdom's consultation on the justification of the Sellafield MOX plant and the United Kingdom's consultation on managing radioactive wastes safely.

Nuclear Entrapment

The process whereby governments can entrap themselves in their commitments, administratively and financially, is examined in a detailed book about THORP that appeared in 1999. William Walker's *Nuclear entrapment: THORP and the politics of commitment* is interesting not only for those who want to know more about the nuclear industry in general, and Sellafield in particular, but for anyone who has ever tried to shift a government away from a strategic or planning decision into which it has begun to plough significant resources. Walker, professor of international relations at St Andrew's University, traces the history of THORP from its origins in the early 1960s and illustrates how political and industrial commitments can lead to entrapment in undesirable technologies and policies, and how such commitments influence democratic institutions.

The Future of THORP

In July 2002, the United Kingdom's White Paper relating to the proposed Liabilities Management Authority (LMA) claimed that, 'Discharges from THORP have a very low environmental impact.' The authors of the White Paper added that, 'Income from the reprocessing contracts being undertaken in the THORP plant is substantial and could make a contribution to clean up costs [at Britain's nuclear facilities].' It is certainly not proposed by the author of the White Paper that the LMA should terminate existing THORP contracts. These will remain with BNFL as site licensee and operator of the plant and will be completed and honoured: 'To do otherwise', wrote the authors, 'would break existing contractual commitments and Government Undertakings. It could also invoke compensation payments which would outweigh the costs involved in meeting those commitments. THORP will, therefore, continue to operate until existing contracts have been completed or the plant is no longer economic.' Indeed, the authors specifically envisaged the possibility of new contracts for reprocessing at THORP. They stated that the United Kingdom government would look in detail at the circumstances of each specific case and 'review the range of issues which would be involved in increasing the current volume of fuel to be reprocessed through THORP', if that arises as a possibility.

The authors of the White Paper noted that, thenceforth, decisions about new contracts for reprocessing at THORP 'would be taken in the best interests of the UK as a whole'. This geographically and politically confined criterion is to be assessed in the light of advice from the Liabilities Management Authority, and on the basis that approval would only be given if the contract was

> consistent with clean up plans for Sellafield and, in the LMA's view, would not cut across implementation of those plans; [and] was expected to make a positive return to the taxpayer after allowing for operational costs, business risks and any other costs which might be incurred as a result of the contract, including any additional clean up costs; and was consistent with the UK's environmental objectives and international obligations.

At the end of November 2002, a sub-committee of the British-Irish Inter-Parliamentary Body expressed 'grave concern' at the failure to return by-products of the reprocessing operation at Sellafield to customers overseas, who are contractually obliged to take them, and expressed the belief that BNFL was actively seeking new reprocessing contracts that would require THORP to be kept in operation beyond 2010.

The extent to which the United Kingdom government had committed itself to THORP and to continued reprocessing was simply underlined by its decision, in 2001, to commission the new MOX plant at Sellafield. It is to that controversial facility that we turn next.

MOX Rocks

There is something almost comic about the ultimate visual destination of a visit to the MOX plant at Sellafield, a plant that has sparked a number of substantial legal exchanges between Ireland and the United Kingdom. Amid all the controversy, behind so much health and security screening, in the presence of plutonium, one finally arrives at a glass window and gazes upon small, dark pellets that look like the droppings of a sheep or goat. But this substance is no joking matter. These little rocks are diplomatic dynamite. Before describing the plant itself, in the following chapter, it is worth contemplating the nature and history of the MOX process itself.

Objections
The word 'MOX', standing for 'mixed oxide', has come also to represent the determination of both the government of the United Kingdom and BNFL to persist with and to expand operations at Sellafield, notwithstanding objections from others. At one point, it had been hoped by some that the major new MOX plant might not open following the discovery that data had been falsified by BNFL at its smaller and older MOX Demonstration Facility at Sellafield. However, those hopes were quickly dashed.

Lies, Damned Lies and Systematic Failure

The safety of an installation such as Sellafield depends on all of its systems being operated in a reliable manner. That is why it is so disturbing to find evidence that data relating to a dangerous process at Sellafield was systematically falsified. In February 2000, the United Kingdom's Nuclear Installations Inspectorate (NII) reported on the falsification of quality assurance data associated with the production of nuclear fuel pellets at the MOX Demonstration Facility (MDF). The particular pellets had been exported to Japan. The British plant was obliged to close for months, while BNFL sought to reassure the authorities that the company could be trusted to continue its operations there. The NII investigation was carried out under the control of the deputy chief inspector responsible for regulating safety at BNFL sites. It began shortly after BNFL notified NII of the suspected falsification on 10 September 1999.

In a striking conclusion to its report, the NII found that,

> It is clear that various individuals were engaged in falsification of important records but a systematic failure allowed it to happen. It has not been possible to establish the motive for this falsification, but the poor ergonomic design of this part of the plant and the tedium of the job seem to have been contributory factors. The lack of adequate supervision has provided the opportunity. Despite this, self-discipline ought to have ensured that those involved followed the proper procedures.

Optimistically, the NII thought that,

> One point worth noting is that in the new Sellafield MOX Plant, currently being commissioned, the inspection processes for MOX pellets, rods and assemblies are designed to be almost fully automated: this should prevent the falsification of data of the kind described in this report.

The latter plant has since been commissioned against the wishes of the Irish government and others. NII did ensure that the MOX Demonstration Facility was shut down and that it was not allowed to restart until the inspectorate was satisfied that the recommendations in its report had been implemented.

The job of those employed at the MOX Demonstration Facility (MDF) at Sellafield was, from the outset, similar in purpose to that of the staff of the newer MOX plant today. It was to manufacture MOX (mixed oxides of plutonium and uranium) fuel pellets for use in nuclear power reactors. On 20 August 1999, a member of MDF's Quality Control Team identified similarities between the 'secondary pellet diameter data' for successive lots of pellets. On 10 September 1999, BNFL reported to the NII that some of these secondary pellet diameter checks on the fuel (manufactured for one of its Japanese customers) appeared to have been falsified by copying some data between spreadsheets. The independent inspectorate promptly launched its investigation to establish both the extent of the falsification and the causes of the event. While the NII found that data had indeed been falsified, the inspectorate concluded that this would not have affected the safety performance of the fuel.

NII described the failure to properly carry out the agreed manual checks of the pellet diameter as 'a deliberate breach of operating procedures'. It added that, 'In a plant with the proper safety culture, the events described in this report could not have happened.' Moreover, NII found that the initial investigation by BNFL, carried out under severe time pressures, was too narrow: 'There had been a tendency to rush to early conclusions which understated the extent of the problem by assuming that the falsification was largely confined to one shift.' One example of falsification was found dating back to 1996: 'The management of the plant allowed this to happen, and since it had been going on for over three years, must share responsibility', concluded the inspectorate.

The MOX Demonstration Facility was permitted to reopen in December 2000. By then, the NII had accepted that various recommendations for new procedures at the Sellafield plant had been implemented by BNFL. One result of the scandal was a serious setback for government plans to attract private investment into BNFL. The MDF facility still operates today, as a support unit for the much larger MOX plant. Environmental groups had vainly hoped that the safety scandal and BNFL's subsequent loss of some reprocessing contracts would mean that the larger MOX plant would never open.

MOX Fuel

MOX fuel is made from a mixture of plutonium and uranium powders and is a valuable product. At Sellafield, BNFL currently makes smooth pellets of MOX fuel for conventional nuclear power stations. The pellets are placed in metal tubes, or 'rods', the ends of which are plugged and welded. The rods are inspected for quality and to ensure that their dimensional characteristics are as desired, before a number of such rods are bolted together in a rectangular 'fuel rod assembly', which is somewhat like a giant version of the element in an electric kettle. These assemblies are later transported to customers by rail or ship.

BNFL describes MOX fuel pellets as 'a hard, ceramic, stone-like material'. The company says that the pellets are so durable that if dropped into water, 'they would take thousands of years to dissolve'. Moreover, adds BNFL,

> The pellets are loaded into fuel rods made from zirconium alloy which are corrosion-resistant and able to withstand depths of several thousand metres of water. The rods are loaded into fuel assemblies which are then loaded into the transport casks. The fuel assemblies are safe enough to allow workers to work immediately next to them.

Up to 200 such assemblies are placed in a nuclear reactor and activated to generate intense heat. Thousands of gallons of water *per minute* are pumped up along these assemblies and create steam for the generation of electricity.

The Sellafield MOX plant has been designed to fabricate this new fuel from a mixture of plutonium, which is recovered from used nuclear fuel reprocessed in the THORP facility, and recovered and re-enriched uranium or 'fresh' uranium. At present, the latter appears to be the preferred option. Captain Malcolm Miller of BNFL told me that,

> In MOX fuel we take plutonium from old fuel, blend it with new uranium, to make new fuel . . . using new uranium from [the BNFL facility at] Springfields. For some reason, we use fresh uranium. Old uranium stored on behalf of customers can be enriched if customers want – and we have done a few shipments of that.

Another BNFL spokesperson pointed out that natural uranium is still being mined 'fairly cheaply', making it economic for BNFL to stockpile its reprocessed uranium for possible future use in the event of political or commercial circumstances changing.

The main markets for MOX fuel are Japanese and European. By reprocessing spent fuel from Britain and abroad, and by keeping the plutonium from it for future use in the manufacture of mixed oxide fuel at the Sellafield MOX plant for clients overseas, BNFL earns money for Britain. It also increases the volume of nuclear wastes and materials being transported internationally. The plutonium that BNFL recovers from reprocessing can be used in conventional reactors or in fast reactors. At the moment, up to one-third of the core of the reactor of a modern nuclear power station can be loaded with MOX fuel. According to BNFL, 'This makes use of about two tonnes of plutonium at any one time. Using MOX helps us to manage the amount of plutonium produced by burning uranium. A tonne of MOX fuel usually contains between about 50 and 70 kilograms of plutonium. After four years of generating electricity, the amount of plutonium in the fuel would have reduced by around 20 per cent.' About 400 tonnes of MOX fuel has been used in conventional nuclear power reactors since 1963. More than thirty European reactors are now licensed to burn it. Future fuel and reactor developments may enable the proportion of MOX fuel loading to increase. BNFL also states that studies by the International Atomic Energy Agency (IAEA) have shown that if MOX fuel is burnt at certain levels, the world's stockpile of plutonium can be held steady and then gradually reduced during the first decade of the next century.

EU Approved MOX
The controversial new Sellafield MOX plant has come on line in a number of stages. Firstly, as a preliminary step, a small-scale facility began working in 1993 to produce fuel for Light Water Reactors. In that same year, 1993, BNFL announced that it intended to build the Sellafield MOX plant.

On 2 August 1996, in accordance with its obligations under Article 37 of the Euratom Treaty, the United Kingdom supplied the European Commission with data relating to the proposed disposal of radioactive waste from the new MOX plant. In fighting recent legal actions by the Irish government, lawyers for the United Kingdom have been eager to point to the decision of the Commission at that time. They have recalled that, on 25 February 1997, the Commission gave its opinion on the MOX plant as follows:

- The distance between the plant and the nearest point on the territory of another Member State, Ireland, is 184 km.
- Under normal operating conditions, the discharge of liquid and gaseous effluents will be small fractions of present authorized limits and will produce an exposure of the population in other Member States that is negligible from the health point of view.
- Low-level solid radioactive waste is to be disposed to the authorized Drigg site operated by BNF plc. Intermediate level wastes are to be stored at the

Sellafield site, pending disposal to an appropriate authorized facility.

- In the event of unplanned discharges of radioactive waste which may follow an accident on the scale considered in the general data, the doses likely to be received by the population in other Member States would not be significant from the health point of view.

In conclusion, the Commission was of the view that the implementation of the plan for the disposal of radioactive wastes arising from the operation of the BNFL Sellafield MOX plant, 'both in normal operation and in the event of an accident of the type and magnitude considered in the general data, is not liable to result in radioactive contamination, significant from the point of view of health, of the water, soil or airspace of another Member State'.

MOX Plant Starts Work

The MOX plant has been 'commissioned' (i.e. gone into production), notwithstanding strong opposition from various objectors to its start-up. As indicated earlier, production at this plant is intrinsically connected to the THORP facility at Sellafield. The Irish government said in October 2001 that it had 'strongly and consistently opposed the commissioning of the MOX plant and its concerns about this plant have been conveyed to the UK authorities in every possible manner and in our responses to each of the five separate rounds of public consultation over the period 1997 to 2001'. During 2000, as seen above, it emerged that, since 1996, certain documents relating to MOX data had been falsified at Sellafield. Moreover, MOX fuel produced by BNFL and loaded into a reactor in Switzerland, ruptured, leaked radiation and had to be removed. However, neither the MOX data scandal nor the faulty fuel was allowed to stand in the way of the Sellafield MOX plant becoming operational.

On 3 October 2001, the United Kingdom government gave its approval to BNFL to use the Sellafield MOX plant (SMP) for

MOX fuel production. The United Kingdom's Department for Environment, Food and Rural Affairs (DEFRA) and its Department of Health had announced that, following the latest period of public consultation into the economics of operating SMP, the plant was justified in their opinion. That same day BNFL stated that the Sellafield MOX plant had already contracted reserved business for the 40 per cent 'break even' sales target, and that there was evidence of customer commitment for much more business besides that. BNFL added that, 'Customers want to recycle their plutonium separated during reprocessing into MOX fuel for use in their reactors. The operation of SMP will enable this MOX fuel to be made in the U.K.' The next day, 4 October 2001, the Irish government said that it saw 'no justification whatsoever for the MOX plant and will do everything possible to bring about a reversal of the UK Government's decision of last week'. On 8 May 2002, BNFL announced that another contract had brought the Sellafield MOX plant close to the 'break even' level of operation.

On 20 December 2001, BNFL commenced the first stage of active plutonium commissioning of the MOX plant. This involved introducing plutonium-bearing material in order to start testing the plant and equipment as a precursor to MOX fuel manufacture. Jack Allen, the new Head of Operations at Sellafield MOX plant, remarked,

> This is wonderful news for SMP and is the best Christmas present we could have had. I am very proud of the MOX workforce who have worked so hard to get us to this stage. I also want to thank our customers who have been very patient and we now want to get on with the job of manufacturing MOX fuel for them. This is just the beginning of MOX fuel manufacture and our focus is now on delivering the first fuel to customers.

The NII had earlier inspected the Sellafield MOX plant, and the Health and Safety Executive (HSE) issued to BNFL the necessary licence consent confirming they were satisfied that BNFL could begin plutonium commissioning of the plant.

MOX Security

The United Kingdom's OCNS was closely involved, for over eight years, in preparations for the coming into operation of the Sellafield MOX plant. Immediately prior to commissioning, further inspections were carried out by the OCNS, including full tests of all security systems and a revalidation of security procedures. The director of the OCNS subsequently wrote in his 2002 report that, 'These showed that the plant met all our security requirements.' Yet, he also acknowledged that in relation to personnel there were 'some outstanding vetting clearances for which temporary compensating arrangements had to be made'. This suggests that those who had access to details of the layout of the MOX plant did not conform fully to security requirements.

Referring in that same report to the attacks of 11 September in the United States, the director 'concluded that commissioning SMP did not increase the vulnerability of the site to this form of attack'. This simple statement could mean that the site was and remains vulnerable. It scarcely reassures those who are concerned about the consequences of an attack. It also begs the question of whether or not nuclear facilities and those who regulate them ought to be obliged to *decrease* the vulnerability of terrorist attack as they expand their own activities on a site, and in the light of the attacks in September 2001. The statement steps around any assessment of the extent to which Sellafield is vulnerable to an air attack, as well as the extent to which the opening of THORP and the MOX plant may have made Sellafield a more attractive target to terrorists and a more dangerous target for the public at large.

Ping Pong

The Japanese were, unsurprisingly, upset when they learnt that eight MOX fuel assemblies sent from Sellafield in 1999

United States following the MOX data falsification scandal, and he is the kind of individual who immediately inspires confidence. His first task today is to shepherd us through even more internal security, which takes almost fifteen minutes. He makes one feel like there is nothing that he would rather be doing, although one guesses that there surely is.

Getting through security at the MOX plant involves yet another BNFL pass, and clearances for cameras. One must enter a special booth, with two doors, swiping a plastic pass and keying in an individual, personal code number. One heavy-set man triggers an alarm that is intended to signal the presence of more than a single body in any entrance booth at the same time. He seems not a bit amused when this happens a couple of times, especially as he himself is an employee of BNFL. But, eventually, they get him through and into the inner world of MOX. There, for us, awaits another change of footwear, with the ritual this time involving steel-capped shoes, and the washing and drying of our hands. All paper towels are carefully directed into bins for special safe disposal, and we are finally asked to put our arms into electronic monitors before being also scanned from head to toe.

By this point, we are becoming accustomed to the constant background sound of the 'Crit. Alarm' sonar bleeping. In the event of a malfunction within the MOX plant, this 'Critical Alarm' will automatically spark an evacuation and other emergency procedures. Occasionally, the electronic bleeping is overridden by a disembodied voice that summons some named employee to a particular place. With their tinny, Tannoy tones, these announcements create the momentary and disconcerting illusion that one is on a normal factory floor. However, at most factories, standard operational procedures do not include special measures for handling plutonium.

The atmosphere in the MOX plant seems very relaxed. It is, apart from the background sound of the 'Crit. Alarm', a quiet kind of place. We enter the MOX control room, which is a spacious area containing a couple of circles and a couple of arcs of control desks, with various computers. Viewers of

The Simpsons might be familiar with this sort of nuclear nerve-centre. But there are no posters of Homer Simpson in here. We are reminded again that photographs may not be taken without prior approval. Once more, the reason for the rule is said to be commercial, rather than related to security.

Talking Point

Jack Allen explains that, in the event of a failure of electricity on site, there is provision for back-up power to allow MOX operators to get the plant 'into a safe position'. In the event of a failure of the back-up power, or other catastrophic occurrence, the whole facility can be closed down manually. The likely result of such a manual shutdown would be major, permanent damage to the MOX plant itself.

Outside the control room is a box containing a sheaf of papers headed 'Weekly talking points'. These single but double-sided A4 pages carry some typed notes about what is happening at the plant. It is the second such box that I have passed, the first being marked 'Please take one' and 'Please take a copy'. But I discover that the invitation is directed only at employees because, when members of the group take them, we are asked to put them back. Especially sensitive, I am told, is the information about Japanese visitors to Sellafield. They do not like publicity, it seems.

Just a Drop

In stark contrast to the THORP facility, with its massive storage ponds, the adjacent MOX plant contains almost no water. This is because water can cause a chain reaction in conjunction with plutonium. The amount of liquid used is, says Jack Allen, 'very, very low' and 'minimal to negligible'. It is 'literally in litres per day', and he adds that the amount of liquid waste from MOX is so small that it is processed in another part of the site where it becomes part of the overall Sellafield discharge.

Allen conveys us in a cavernous, stainless steel elevator to a floor where we can see some MOX pellets in glass cases.

There are special armholes and rubbery gloves built into the glass, allowing Sellafield employees to manipulate and treat the pellets in various ways. As said earlier, the small pellets themselves look somewhat like shiny sheep-droppings. Those whose hands might otherwise be found in the rubber gloves at work on the pellets appear to be out to lunch. The room has the deserted air of a dusty science laboratory at an old university, during the summer vacation. But it is, no doubt, at the very cutting edge of British technology.

When we have had our fill of looking at the dull and stationary objects in the room with pellets, it's down again in that big, shiny steel lift to the MOX Export Facility, to see where new MOX fuel is placed in cylinders for despatch to BNFL customers. Such fuel is slightly warm, giving off some radiation, but it is not technically 'critical' at this stage. We are not invited to feel the heat. Jack Allen reminds us that the MOX plant is making fuel for Pressurised Water Reactors and Boiling Water Reactors. It is not making fuel for Advanced Gas-Cooled Reactors or Magnox Reactors, he adds. Pressurised Gas-Cooled Reactors are the most common reactors worldwide, with about 400 internationally, but Sizewell B is the only one in Britain. In fact, MOX depends for its existence on the foreign market. Allen, the American, is reluctant to comment on Britain's utilisation of the Magnox and AGR types of reactor. Do I perceive a twinkle in his eye as he compliments the United Kingdom on its standards and initiatives in the 'early' phases of nuclear power?

That's It?
There is not a lot to see at the MOX plant besides walls, containers, control panels and some pellets. Clean, hard surfaces abound, but the most important feature is invisible. It is radioactivity. The most interesting substance that emits radioactivity at Sellafield, plutonium, is itself not something that is visually distinctive. So nodding to a few preoccupied MOX staff, who pass occasionally like PNTL ships in the night, I conclude that I have 'done' the MOX plant.

In my ignorance, before leaving the building, I ask one senior employee if he finds it boring to work there. After all, as may be recalled from an earlier chapter, 'tedium of the job' was officially identified as a factor contributing to the falsification scandal of 1999. The new MOX plant seems so perfectly automated that one wonders how many ways there can be to make or monitor a MOX pellet. The senior employee whom I have interrogated indicates politely that I have not the slightest idea of the diversity and extent of the challenges that face him in the course of his day's work.

Contortions

Leaving the MOX plant behind me, I return through the THORP complex to recover my own shoes. Again, I pass through layers of security and monitoring. The penultimate monitors malfunction, due to their sensitivity to the light from fluorescent bulbs in the inner changing room at THORP. It appears to have happened before, to be a problem to which they are accustomed, and we shrug and move on. The final radiation monitor is a booth, where one engages in various contortions. I push my arms into metal holes, while thrusting forward my chest and chin against a steel panel. Then I must hold my breath until a bleep sounds, then turn and press backwards hard while simultaneously reaching out and pressing two big knobs on the opposite wall of the booth, until the outer door opens.

Finally, I am reminded to bin the blue, woolly socks that I had been instructed to pull on over my own socks at the start of the visit to THORP and MOX. I do so, dutifully, before exiting THORP through one last security screen. It is good to step into sunlight and to feel a breeze on my face, even if it is blowing across a big flask of nuclear fuel that sits on a railway carriage a few yards in front of me.

MOX Lunch

Whisked off the Sellafield site, passing quickly over the unassuming pipes that snake down a ditch to deposit radioactive waste in the Irish Sea, we are brought back a short distance along the road to Pelham House. This beautifully restored mansion now houses some of the administrative offices of BNFL. Here, we again meet Jack Allen, who now hosts a buffet lunch for us. Is there no end to this man's politeness? He is eager to pour water or juice for his guests, in an appealing and unreserved American way. He is also anxious to let me know just how committed he and his staff are to high standards of care and maintenance at the MOX plant.

Allen acknowledges that the MOX data falsification scandal, which occurred before he arrived, damaged BNFL. He says that the staff involved have either left or been moved elsewhere. He praises the local Cumbrian people whose level of general education and knowledge about the plant contrasts very favourably with that of the population around the nuclear plant in Carolina, USA, at which he formerly worked. After dessert, which consists of a strange but refreshing ice-cream cake that is a Sellafield speciality, we bid one another farewell.

The Future of MOX

In July 2002, the authors of the British White Paper on nuclear liabilities management signalled that the Sellafield MOX plant (SMP) has an active future, at least so far as the government of the United Kingdom is concerned. They observed that,

> The economic case for operation of SMP was carefully considered by the Government, informed by advice from independent expert consultants. It was based on a prudent assessment of likely sales of MOX fuel to Japan, Germany, Switzerland and Sweden, using the plutonium arising from existing spent fuel reprocessing contracts. As was made clear in the 3 October 2001 decision by the Secretaries of State for Health and for the Environment, Food and Rural Affairs on the

justification of MOX manufacture, the economic case was demonstrated to be strongly positive compared to the non-operation of the plant.

The White Paper claimed that,

SMP's operation does not generate substantial new issues in terms of decommissioning and waste management and produces insignificant discharges of radioactivity to the environment.

The authors wrote that BNFL would continue to seek contracts for the supply of MOX and that, even if and when any major restructuring of BNFL takes place, 'existing contracts will remain with BNFL plc as site licensee and operator of the plants and will be honoured'. It is intended that the 'New BNFL' will continue to be responsible for all dealings with THORP and SMP customers. The arrangement is meant to enable the new Liabilities Management Authority to focus on its responsibilities to clean up Britain's older nuclear sites. However, decisions on the future of THORP and SMP will be taken by the Secretary of State on the basis of advice from the LMA, and not ultimately from BNFL.

Walking through the extremely expensive buildings which house THORP and MOX, and reading recent British reports on various aspects of the nuclear industry, one quickly forms a distinct impression that it is highly unlikely that the government of the United Kingdom will bow to demands that it close down its spanking new civilian nuclear facilities at Sellafield.

CHAPTER 14

Bright Hopes, Dark Cloud

S ellafield is likely to be with us for quite some time, in some shape or form. There are environmental, commercial and strategic arguments in favour of sustaining the international nuclear industry. These arguments are regarded as persuasive by a number of governments and may soon be strong enough to result in new nuclear reactors being built, even in European countries where there has been an effective moratorium on their construction. The nuclear industry may triumph, notwithstanding the dangerous legacies of earlier nuclear policies, and despite its dark past. Significantly, in January 2002, the government of Finland made a favourable decision 'in principle' on an application to build a fifth nuclear power plant in that country.

The Environmental Argument
The Kyoto Protocol, which is an international treaty aimed at restricting certain emissions of 'greenhouse gases' that harm the earth's atmosphere, and for which environmentalists fought hard, creates serious challenges for energy companies that burn fossil fuels such as oil and coal. But Kyoto has created a new opportunity for the nuclear industry. When nuclear power plants run well and safely, they are relatively clean and efficient compared to fossil fuel plants.

There is a rapidly growing demand for electricity throughout the world. All methods of generating it have some impact on the environment. For example, burning coal releases big quantities of carbon dioxide into the atmosphere. However, greenhouse gases and acid rain, which are associated with fossil fuels, are not a problem when nuclear power is generated. It has been estimated that in 1999, for example, nuclear generation saved the use of around 264 million tonnes of coal, 1,517 billion cubic feet of natural gas and 66 million barrels of oil worldwide. It is claimed that, globally, electricity supplied by nuclear power stations avoids the emission of around two billion tonnes of carbon dioxide annually. In 2001, many countries, including Germany, India, Spain, Switzerland and the United States, set record levels of production of electricity from nuclear power.

International efforts to control carbon dioxide emissions would be adversely affected if nuclear power plants were to be shut down. The United Kingdom's Department of Trade and Industry has publicly acknowledged the role that nuclear power plays in limiting such emissions. Within the European Union many countries use nuclear power, and the EU Commission has recognised the contribution that the nuclear industry makes to helping to meet the commitments under the Kyoto Protocol. Last year, approximately one-third of Europe's electricity was generated by nuclear power.

Globally, the International Atomic Energy Agency (IAEA) serves as the world's foremost intergovernmental forum for scientific and technical co-operation in the peaceful use of nuclear technology, including nuclear power. Established as an autonomous organisation under the United Nations in 1957, and based in Vienna, the agency controls and develops the use of atomic energy. It estimates that there were 438 nuclear power plants in operation at the beginning of 2002, and points out that, while nuclear power is being generated primarily in a broad range of industrialised countries, 31 of 32 new plants under construction are in Asia or in Central or Eastern Europe. The IAEA notes that about 16 per cent of the world's electricity is generated by nuclear power.

In this context, attempts by Irish people to wage a campaign for the closure of one particular nuclear facility in Britain are unlikely to win universal support. The European Commission is believed to have been displeased by the Irish government launching a number of legal actions against Sellafield. Ireland has been eager not to appear mischievous and actually couches its legal arguments for safety and accountability in the context of the peaceful promotion of nuclear energy, which is facilitated by various international agreements and agencies. Ireland's case is anti-Sellafield rather than anti-nuclear.

At the Council of Environment Ministers, held in June 2002 in Luxembourg, Ireland's then senior Minister for the Environment, Noel Dempsey, attempted to ensure that the European Union's negotiating position in forthcoming climate change negotiations would resist support for nuclear power. The ministers had assembled to agree their stance for a forthcoming series of negotiations on implementing the Kyoto Protocol on reducing emissions of greenhouse gases. The Irish government noted that the European Union faces pressure to agree the construction of new nuclear power stations in developing countries. The Kyoto Protocol provides for a number of mechanisms to stimulate investment in greenhouse gas reduction technologies.

In the discussions between European Union ministers, the Irish argued that, 'the EU should not try to solve one environmental problem (climate change) by endorsing or supporting technology (nuclear) which creates additional and intractable environmental problems, and has the potential to be a very grave risk to human health and safety'.

The Commercial Argument
In its most recent general election manifesto, the British Labour Party looked to expanding rather than contracting the operations of BNFL. The party proclaimed that, 'BNFL is an important employer and major exporter. The Government insists it maintains the highest health, safety and

environmental standards. We are examining the scope for turning the company into a Public Private Partnership.'

In the United States, nuclear energy has been placed firmly on the agenda by President George W. Bush. Generally regarded as a close ally of the energy industry, which is home to many of his strongest backers and associates, Bush is not opposed to expanding the provision of nuclear power.

In Britain, the planned closure of older nuclear power stations and the economic difficulties encountered by British Energy in managing its more modern ones, has concentrated the minds of those in government on the question of how Britain is to sustain its own energy needs. Most existing British nuclear power plants are due to be decommissioned between now and 2025. At the time of their peak output, in the 1990s, around 30 per cent of all electricity used in Britain came from nuclear sources. Since then, the proportion has declined somewhat and is expected to fall to about 7 per cent or 8 per cent by 2020, although it may be noted that most of Scotland's power still comes from nuclear sources. The UK government has undertaken a fundamental review of energy policy, which raised the possibility of new nuclear power plants being built in Britain. These would be most likely to be on existing nuclear sites. However, the UK government has denied that the creation of a new Liabilities Management Authority to care for older nuclear facilities is a backdoor route to the creation of new sources of nuclear power. Debate continues over the 2003 White Paper on Energy.

Meanwhile, BNFL has been busy expanding internationally. It offers its reprocessing services at Sellafield to foreign clients, transporting nuclear wastes to and from Britain by ship. BNFL has also made a number of important international acquisitions, most notably that of the Westinghouse group, which owns nuclear power plants across the United States. These are not the activities of a company that is gearing up to go out of business.

Even if Britain were entirely and permanently to discontinue the nuclear generation of electricity, and there has been, since the 1990s, a moratorium on the building of new nuclear power plants in the United Kingdom, the reprocessing plant at Sellafield could probably long remain operational. Reprocessing spent fuel and storing waste from existing stations at home and abroad is a source of potentially massive earnings for the economy of the United Kingdom. Moreover, the nuclear industry's role as a centre for research relevant to military and energy needs should not be overlooked.

Because Sellafield itself is so crucial to the British nuclear effort, and is a major source of local employment in Cumbria and of potential national revenue, the government of the United Kingdom is unlikely to close the plant entirely in the foreseeable future. It has spent enormous sums of money on developing it, and the government is advised by its officials that the plant is safe.

The Strategic Argument

Closing Sellafield could also have implications for the warm strategic relationship between Britain and the United States. Tony Blair could be thought to be weakening the case for more energy at a time when George W. Bush is forging ahead with plans to expand nuclear production at home and oil drilling abroad. To close off the nuclear power option, when alternatives such as wind and solar power are still not regarded by governments as an adequate substitute source of electricity, could leave Western countries even more dependant than at present on oil from the Middle East. Recent developments in Afghanistan and Iraq have served to remind people of the potential problems of depending on energy supplies from that region.

Reprocessing is attractive because it reduces dependence on uranium supplies also. As BNFL put it in a briefing note for journalists,

> Another reason for opting for reprocessing is strategic. The oil crises of the 1960s and 70s showed that energy prices are not only fixed by supply and

demand, but by politics and the state of trading. At the moment there is plenty of uranium, but politics could affect how much is made available. Reprocessing should help keep uranium prices stable by providing an alternative to mining. A number of countries have decided to use reprocessing as part of a long-term policy.

So long as the option of nuclear power is kept open on both sides of the Atlantic, then the operational future of any reprocessing and storage facilities at Sellafield seems all but guaranteed. Its closure, while welcome news for some environmentalists and others, could render the global nuclear power sector less stable. Governments tend to fear a domino effect, suspecting that if they concede to demands for the closure of one plant, then it will become more difficult to resist calls for the closure of others.

The special relationship between Britain and the United States takes various forms. During 2002, it emerged that subsidiaries of BNFL, which owns Sellafield, had made financial donations of hundreds of thousands of dollars to American politicians and political groups. These donations were not clearly revealed in BNFL's United Kingdom accounts and are said by BNFL to have been made without the knowledge of BNFL's chief executive, Norman Askew. When they became public knowledge, it was stated by BNFL that they would not be repeated. The payments underline the importance of politics in determining the future of energy corporations.

Also, in July 2002, the chief executive of BNFL, Norman Askew, said that, 'it is very pleasing to see that the Congress of the United States has agreed to proceed towards the commissioning of Yucca Mountain as a long-term storage facility for spent fuel. This and other initiatives being taken in the USA is paving the way for new nuclear build [i.e. new power plants].'

Euratom
Governments of particular European countries have for some time expressed reservations about, or even opposition to, nuclear power. Nevertheless, as indicated above, European policy remains fundamentally disposed towards the development of nuclear energy, in accordance with the 'Euratom' agreement.

The European Atomic Energy Community (Euratom) is an organisation that was established by treaty in 1958 to create the conditions necessary for the establishment and growth of nuclear industries. The United States promoted its establishment to facilitate the sale in Europe of US nuclear power reactors and related fuels equipment and technology. In time, Euratom grew to have a nuclear technological base rivalling that of the United States itself. Many Member States of the European Union have commercial nuclear power plants, and several rely heavily on nuclear power. Euratom members are parties to the Nuclear Non-Proliferation Treaty and belong to the International Atomic Energy Agency.

Two fundamental objectives of the Euratom Treaty are to ensure the establishment of the basic installations necessary for the development of nuclear energy, and to ensure that all users in its community receive a regular and equitable supply of ores and nuclear fuels. The Euratom Supply Agency, operative since 1960, is the body established by the Euratom Treaty to ensure this supply. The Euratom Supply Agency acts under the supervision of the European Commission.

Euratom is something of a potential embarrassment for Ireland. This is because the agreement of *all* European Union Member States is required to increase its budget. Ireland adopts a ploy which enables it to remain publicly opposed to the expansion of nuclear power while not appearing to be obstructive to European policy. Thus, the Irish government supports the extension of Euratom loan facilities to Eastern Europe, for example, on the basis that the loans are to be used to enhance the safety of the nuclear industry there. However, critics claim that such 'conditions' are unenforceable in practice, and that such loans may even be used to build reprocessing facilities similar to those at Sellafield. For one particular critique of Ireland's position on Euratom see http://www.eu-energy.com/Euratom.html.

Ready To Cash In
BNFL, proud owner of Sellafield, is ready to cash in on current efforts to protect the environment, and on the reorganisation of the nuclear industry in Eastern Europe and elsewhere. In his annual report for 2001, BNFL's chairman, Hugh R. Collum, described the attitude of Britain's Labour Party to the state-owned company as 'encouraging'. He noted that 'the world is changing', and thought that, 'with the longer term picture for the global nuclear industry brightening, the future for the business is encouraging'. He continued,

> At the end of last year I called for an informed and open public debate on the future of nuclear energy. There are clear signs that this is underway as governments, politicians and public opinion recognise the important role that the nuclear industry can play in meeting the world's demands for long-term energy supply as well as addressing the environmental issues.

The chairman of BNFL is heartened by developments across the Atlantic. He wrote that,

> In the United States particularly, there is a new urgency to address environmental and energy shortage issues through a broader reliance on clean, safe nuclear power in the future. This was strongly demonstrated in May with President Bush's energy policy announcement, which puts nuclear firmly back on the agenda.

Collum was referring here to the new US 'Energy Policy' released in May 2001, which recommended government support for 'the expansion of nuclear energy in the United States as a major component . . . of national energy policy'.

The BNFL chairman next turned to the United Kingdom itself, where the closure of all but one nuclear station by the year 2025, as planned, would result in an energy mix in which nuclear plants would have only around a 3 per cent share of the United Kingdom's electricity supply market. He claimed that, environmentally, if the output from nuclear plants were to be replaced by conventional power stations, then 'much of the hard-won carbon dioxide emission reductions achieved so far would be lost'.

Irish Unimpressed

Speaking in Seanad Éireann (the Irish senate) on 10 October 2001, the minister responsible for Irish nuclear policy, Joe Jacob, made it clear that his government does not share bullish sentiments about the prospects of growth in the nuclear power sector. Jacob told his fellow parliamentarians that the future of nuclear power worldwide remains 'highly uncertain'. He observed that some countries have decided to reject nuclear power but admitted that,

> On the other hand, arising from concerns about global warming, climate change and the need for sustainable development in energy, Ireland is acutely aware that some governments, international organisations and other stakeholders see these concerns as an appropriate basis on which to re-launch and re-invigorate the nuclear solution as a response measure.

He said that,

> Ireland is firmly of the view that nuclear energy is incompatible with the objectives of sustainable development principally because of the real risks which nuclear energy continues to present in terms of security and safety, such as the transport of nuclear materials, radioactive waste and spent fuel management, environmental contamination and increased proliferation risks.

Jacob told senators that the Irish government believes that the suggestion that nuclear technology might be a solution to solving the problem of greenhouse gas emissions does not hold merit, because of the dangers involved. He said that countries which have rejected nuclear energy should not be exposed to such risks and that,

> It is our firm view that the risks posed by nuclear energy should always mean that it has no role to play in any sustainable development policy.

He added that,

> Ireland would be concerned if developing countries were encouraged to develop nuclear industries as part of their energy mix and their efforts to strengthen their economies. It is Ireland's view that it would be a mistake to encourage this approach.

Britain's Evolving Energy Policy

The government of the United Kingdom has been considering the option of constructing new nuclear power stations, although none were built in recent years and those still operational are scheduled to close in the coming years. However, the fact that the UK government was entertaining the possibility of new nuclear power plants became obvious when a report on Britain's energy needs was published in February 2002 by the Performance and Innovation Unit [PIU] of the Cabinet Office. The full PIU report is available on-line at www.piu.gov.uk. This report recommended explicitly that the United Kingdom's Department of Trade and Industry 'should take the necessary actions to keep the nuclear option open'. The Cabinet Office unit noted that, 'The Californian crisis [when power cuts resulted from an excess of demand over supply] has highlighted the importance of putting in place the right incentives for investment in energy infrastructure. And the UK is likely to face increasingly demanding greenhouse gas reduction targets as a result of international action, which will not be achieved through commercial decisions alone.'

The Cabinet Office unit described the introduction of liberalised and competitive energy markets in the UK as 'a success', which it felt should provide a cornerstone of future policy. But the 'success' of competition in driving down prices for their consumers of electricity also made the expensive option of nuclear investment less attractive for private financiers. The Cabinet Office unit observed, pointedly, that 'new challenges require new policies'. The unit proclaimed that 'Key policy principles should be: to create and to keep open options to meet future challenges; to avoid locking prematurely into options that may prove costly; and to maintain flexibility in the face of uncertainty.' It noted that, 'Increasingly, policy towards energy security, technological innovation and climate change will be pursued in a global arena, as part of an international effort.' In this context, it is notable that current EU and US policies towards nuclear power appear to be more overtly favourable than that of the United Kingdom in recent years.

The report from the Cabinet Office of February 2002 was followed, just three months later, by the publication of an inter-departmental consultation document. Entitled *Energy policy: key issues for consultation*, this was a further step towards producing a White Paper (official proposals) on the subject and may be downloaded at http://www.dti.gov.uk/energy/developep/index.htm. It echoed the sentiments of the Cabinet Office report, thus keeping open the nuclear option.

Then, in September 2002, the government of the United Kingdom stepped in to rescue British Energy, the ailing private company, and its formerly publicly owned nuclear power stations. The government indicated that it had no intention of letting nuclear power wither away simply because new electricity trading arrangements had, for the time being, made its generation unattractive to investors.

Dark Cloud

Whether or not it decides ultimately to proceed to commission new nuclear power plants in Britain, or even to shut Sellafield, the government of the United Kingdom must make definite decisions about the management and decommissioning of its present, ageing nuclear facilities. It must also solve the long-term problem of disposing safely of existing and future nuclear waste. The problem of nuclear waste is closely related to the functions and future of Sellafield and is a major issue for both the British and Irish governments. During 2002, the government of the United States decided that Yucca Mountain, Nevada, ought to become that country's long-term storage facility for highly dangerous nuclear waste. Attempts to select a site in Britain for the nuclear waste of the United Kingdom have so far been unsuccessful. The issue of waste storage will be considered in the next chapter.

And, regardless of how the problem of indefinitely storing radioactive waste is resolved in the long term, if indeed it can be resolved satisfactorily, there is a more immediate difficulty. This is the challenge of simply maintaining present nuclear facilities and their contaminated and radioactive contents, as well as the waste that has been stored 'short-term' on existing sites.

Managing Liabilities

As the nuclear industry has aged, it has created a growing number of liabilities for the future. Each facility remains operational for a certain period of years, but it does not cease to be a liability when it ceases to be operational. Already, there are many buildings and sites that are a financial burden on the industry, generating no revenue but costing huge amounts of money to maintain safely for an indefinite number of years. These need to be 'decommissioned', and their contaminated parts stored long-term with other dangerous products of the nuclear industry. Such liabilities threaten to make the nuclear energy sector highly unattractive to investors. It is feared by some observers that not enough cash is being set aside to take care of the nuclear legacy in future years.

Responding to these circumstances, on 28 November 2001, the United Kingdom Secretary of State for Trade and Industry announced a proposal to set up a Liabilities Management Authority (LMA). In an earlier chapter we saw that the creation of an LMA will free BNFL to pursue particular investment opportunities. The Secretary of State has indicated that the authority is to take over all of the nuclear sites and facilities now operated by the United Kingdom Atomic Energy Authority (UKAEA) and BNFL that were developed in the 1940s, 1950s and 1960s to support the government's research programmes, as well as the wastes, materials and spent fuel produced by those programmes. It will also take over BNFL's Magnox nuclear power stations, the last of which will cease to be operational before the end of this decade. Finally, it will take on responsibility for the Sellafield reprocessing plant, with its accumulated wastes other than those belonging to customers who have contracted to take back their waste in due course (such as the overseas customers for reprocessing at THORP). These combined legacies are said to represent about 85 per cent of total nuclear liabilities in the United Kingdom and are at present, in any event, wholly the responsibility of the United Kingdom government. Thus, the LMA will take over direct financial responsibility for all of the liabilities that BNFL has formerly managed except those covered by certain commercial contracts.

In announcing the government's intention to create a Liabilities Management Authority, the Secretary of State said that he saw the body as 'providing the driving force and incentives to get on with the job of systematically and progressively reducing the hazard posed by legacy facilities and wastes. It will have a specific remit to develop an overall UK strategy for decommissioning and clean-up.' The establishment of the Liabilities Management Authority and the transfer of assets and liabilities from BNFL and the UKAEA to it requires primary legislation. Among other things, says the director of the OCNS, 'We shall need to ensure, in particular, that any new management structures or operators brought in by the LMA can continue to deliver effective security management in response to regulatory requirements.'

Whitewash?

In July 2002, the United Kingdom government proceeded to publish a White Paper giving further details of the proposed Liabilities Management Authority. This is a detailed and useful document for anyone concerned about the implications of the world's nuclear past for the world's future. However, Ireland's Minister for the Environment, Martin Cullen, responded to the appearance of the White Paper by observing that, while establishment of the new authority might lead to an improvement in nuclear waste management in Britain, 'it would effectively allow BNFL to operate independently of the loss-making side of its business and concentrate more on its core activities including reprocessing of spent nuclear fuel'. Accordingly, added the minister, 'the establishment of this Authority does not in any way reduce Ireland's concerns about the Sellafield operations. The Irish Government remains steadfastly opposed to the Sellafield operations and will continue to press for the cessation of all operations at the plant.'

Moreover, it is not a simple task to separate the legacy of Britain's nuclear past from the future management of its operational aspects. Thus, the authors of the recent White Paper on liabilities management themselves observe that, 'BNFL operates a range of plants and facilities at its Sellafield site, in particular THORP and SMP [Sellafield MOX plant], providing commercial services to private sector and overseas customers. These, and the wastes, materials and spent fuel at Sellafield owned by BNFL's commercial customers, are not part of the legacy.' Attempts to define Britain's nuclear legacy do not necessarily result in neat operational solutions to all matters of waste handling, as the authors of the White Paper seemed to realise: 'THORP and SMP were built with decommissioning in mind and do not present the problems associated with legacy plants. However, whilst commercial customers will retain ownership of their wastes, the integrated nature of the Sellafield site is such that, for regulatory and managerial reasons, legacy and commercial activities have to be treated as a single whole.'

Nevertheless, the bottom line appears to be that the UK government wants BNFL, henceforth, to operate largely as a management company, freed of many of its existing burdens of caring for old plant and waste. That expensive and dirtiest of work will become the responsibility of the state's Liabilities Management Authority. It is a change that, of itself, may not alter the nature or burden of Britain's nuclear legacy one whit. Its precise implications for the safe management of Sellafield are not yet clear.

What is evident is that the early enthusiasts for nuclear research did not envisage the scale or complexity of the problems that they were bequeathing to future generations.

CHAPTER 15

A Poisoned Chalice

The ancient Sellapark House, where guests of BNFL are lodged at Sellafield, is reputedly haunted. Even before hearing about its ghost, I had experienced a strange sensation in my room there. Late at night, the atmosphere seemed to grow colder and darker, and the thought flickered across my mind that this was just the sort of place that might harbour a malevolent spirit. It is doubtful if many of BNFL's scientists believe that ghosts are anything other than figments of the imagination, but they cannot deny the fact that the nuclear industry itself is haunted by its past. Radioactive wastes, accumulating down the years, have changed the way in which nuclear power is perceived.

Nuclear wastes arise from many different sources, including the manufacture of nuclear fuels and the operation and decommissioning of nuclear power stations. They also come from military activities, from nuclear research and from medical, industrial and other uses of radioisotopes. Much of Britain's civilian nuclear waste is stored in and around Sellafield. Some radioactive waste from Sellafield itself is also deliberately poured into the Irish Sea, ostensibly in accordance with the terms of its licences. What is said to be the most dangerous waste at Sellafield has yet to be found a 'permanent' home. It is a poisoned chalice that must be handed down from generation to generation. The consequences of the contents of that chalice being spilt or vaporised, unlikely as such an eventuality is said to be, could be catastrophic.

The disposal of all waste presents society with long-term problems. Traditional rubbish dumps can be a hazard to our health and to our environment, when a wide range of unwholesome objects are poured into holes in the ground and covered over. Fridges and plastic, among other materials, present particular problems of disposal. Some industries produce large quantities of hazardous chemical waste. It is misleading to single out radioactive waste as though it alone is a problem. However, radioactive waste can be extremely and especially dangerous for humanity, and its long or indefinite active life means that its existence will remain a problem for many generations to come. It may not constitute a large proportion of our rubbish, statistically, but it is certainly a very significant part of it.

According to the United Kingdom's Department of Trade and Industry, 'Over four million cubic metres of toxic waste is produced in the UK each year, while radioactive waste amounts to only about 1 per cent of this.' Although there is no long-term storage facility for radioactive waste in the United Kingdom, the government of the United States has recently designated Yucca Mountain in Nevada as the proposed long-term storage facility for US radioactive waste. In the absence of such long-term facilities, nuclear wastes are being kept in various conditions and at a variety of locations.

Responsibility

The owners of radioactive waste are responsible for dealing with future waste 'arisings', as well as with those wastes which already exist. The extent to which they have made provision, in practice, for the care and safe storage of their waste is a matter of debate. For example, how does one calculate precisely how much money will be needed in the future to store nuclear products indefinitely? It is clear, already, that early assessments of what was adequate in terms of future nuclear liabilities were mistaken. Critics also worry that the financial imperative of maintaining a nuclear business mitigates against the setting aside of adequate resources for caring for waste in the long term. When economically and socially justifying the construction of new nuclear facilities, companies and governments may be tempted to underestimate their ultimate true cost to themselves and to society.

The stated policy of the government of the United Kingdom is to ensure that radioactive waste is managed safely and that the present generation meets its responsibilities to future generations. Nuclear wastes are classified according to their level of radioactivity (low, intermediate or high), and their form (solid, liquid or gaseous). There are various ways of disposing of them according to these two classifications. Much waste has to be stored in cooling ponds and in certain long-term facilities. The nuclear industry is accumulating an inventory of such radioactive waste, including spent fuel and the infrastructure of old power plants that have been decommissioned because of age or other factors. Much nuclear waste will continue to be dangerous long after we and our children are all dead. The reprocessing of used or 'spent' nuclear fuel itself results in the production of certain useless waste components. These radioactive by-products emerge as waste streams, and also must be stored carefully.

Long-Term Problems
In September 2001, the United Kingdom's Department for Environment, Food and Rural Affairs, in association with a number of other government departments, published *Managing radioactive waste safely: proposals for developing a policy for managing solid radioactive waste in the UK*. This discussion document on the future management of solid radioactive waste, a ninety-one-page document, is an essential text for anyone concerned about the problem. The paper contains some stark statements relating to 'solid' radioactive waste. Solid waste includes parts of decommissioned power plants, but excludes dangerous liquid wastes that have not been 'vitrified'. The paper states that,

More than 10,000 [metric] tonnes of radioactive waste are safely stored in the U.K., but await a decision on their long-term future. This will increase to 250,000 tonnes when nuclear material currently in use is converted into solid waste. Even if no new nuclear power plants are built and reprocessing of spent nuclear fuel ends when existing plants reach the end of their working lives, about another 250,000 tonnes of waste will arise during the clean-up of those plants over the next century. Most of this waste results from the work of government agencies or publicly owned companies since the 1940s. Some of the substances involved will be radioactive and potentially dangerous for hundreds of thousands of years.

In addition, note the authors, there are much larger amounts of low-level radioactive waste, which is currently being disposed of mainly near Sellafield, at the Drigg facility (which will be discussed later below). Copies of the discussion document are available on the DEFRA website: www.defra.gov.uk/environment/index.htm.

Billions of Pounds
Certain plutonium which is unsuitable for recycling in MOX fuel constitutes a significant part of the growing volume of nuclear waste. Whether or not new British nuclear power stations are constructed that might use up more plutonium in reprocessed fuel, massive and costly challenges remain for those charged with looking after it safely. The authors of the 2001 report on managing Britain's waste considered various options for the storage of plutonium, and their conclusion is staggering. They point out that if the option to re-use were not taken, the material could be treated in readiness for its long-term management. But, they say,

Even on the assumption that such action would not take place for many years, the requirement for such provisions would have significant financial implications for the owners of the plutonium. As the main owners of the material, the majority of the liability would fall to the public sector. Accurate costings for treatment and possible long-term management options for plutonium are not available and estimates vary widely. *They are however likely to be of the order of billions of pounds* [italics added by the present author].

The government of the United Kingdom has appointed a Radioactive Waste Management Advisory Committee. Presumably, it is the 'waste' in the title of this august body, and not the 'committee' itself, that is 'radioactive'. Or, perhaps, they have been exposed to something we don't know about.

Waste at Sellafield
There have been periodic, official reviews of how the radioactive detritus produced at Sellafield is processed and stored by BNFL. In 1996, for example, the United Kingdom's Environment Agency and its Health and Safety Executive jointly published a report on the management of liquid waste and sludge that had accumulated at Sellafield. Stored intermediate-level radioactive waste at Sellafield was found to consist of concentrated aqueous liquors and spent organic solvent. The sludge had arisen in fuel storage ponds. Investigators felt that there was scope for improving the treatment of pond-water to inhibit algae growth that is a significant source of sludge. They recommended improvements, including the management of a redundant low-level radioactive liquid effluent treatment facility. They also found that a relatively large amount of potentially contaminated waste lubricating oil had been accumulating on the Sellafield site as a result of routine maintenance for many years.

But the waste at Sellafield which gives rise to the greatest concern is that stored in the facility known as 'B215', which has been mentioned earlier and which shall be considered further below.

Since 1976, contracts with outside customers for the reprocessing of fuel at Sellafield have made provision for the return to customers of some of the waste produced. When it arrives at Sellafield, such fuel is first stored underwater in the Fuel Handling Plant, opened in 1985 by Prime Minister Margaret Thatcher.

At Sellafield and elsewhere, radioactive waste is held in anticipation of the eventual creation of an alternative and satisfactory long-term dump ('repository'). In this context, it is intended that the proposed new Liabilities Management Authority will take on the responsibilities for the waste being held prior to its long-term storage.

The three main levels of nuclear waste, being high, intermediate and low, are known respectively as 'HLW', 'ILW' and 'LLW'.

Sellafield's High-Level Waste (B215)
High-level waste (HLW) is the most radioactive form of waste produced by the nuclear power process. It consists, principally, of materials separated from uranium and plutonium during reprocessing. For safer storage, liquid high-level waste can be turned into a solid glass form ('vitrified'), within stainless steel containers. The process of 'vitrification' reduces the volume of waste by two-thirds. The challenge of safely storing high-level waste into the future, even for millennia, is currently exercising official minds.

According to the United Kingdom's Department of Trade and Industry, 'High-level, or heat generating, waste which constitutes about 0.3 per cent by volume of all nuclear wastes, arises only from the reprocessing of spent nuclear fuel. It is stored in raw form at Sellafield and Dounreay in special tanks. It is being converted into vitrified (glass) form at Sellafield where it will be stored in purpose built facilities for at least 50 years to cool before disposal.' Thus, highly radioactive liquid waste can be made safer by turning it into a form of solid glass or ceramic, through the process known as 'vitrification'. Critics of Sellafield want that process speeded up.

According to BNFL, there are still about 1,500 cubic metres of high level, radioactive liquid waste at Sellafield. As mentioned above, this waste, or 'liquor', is kept in the complex of storage facilities at Sellafield known as B215. Also stored at B215 is the 'glass' into which some liquid waste has eventually been solidified. B215 has evolved into its present form over a period of forty-four years. The facility has been maintained in a way that is expected to withstand earthquakes, unusual winds and other meteorological conditions. Its features have been re-evaluated by BNFL since 11 September 2001. The company asserts that B215 is unlikely to suffer catastrophic damage even if it is hit at high speed by a civilian aircraft with a full load of aviation fuel.

However, referring to B215 in February 2000, the Nuclear Installations Inspectorate of the HSE reported that, 'There has been no specific design provision against crashing aircraft.' This was because 'the effects of aircraft crashes have been assessed and BNFL concludes that in absolute terms the likelihood of an aircraft impact onto any individual plant is very remote (i.e. the total impact is below 1×10^{-6}/year)'. Indeed, not all crashes at Sellafield would necessarily have radiological consequences. The NII added that, 'Impacts capable of perforating the massive secondary containment have been computed at a frequency of 1.7×10^{-8} or less.' That same report contains considerable further details of how B215 had evolved and been operated, up to the year 2000. The HSE admitted that, 'the safety of the storage of highly active liquor (HAL) has consistently been one of the more significant public concerns associated with the reprocessing of irradiated nuclear fuel and the management of radioactive wastes. This was evident at the Windscale Inquiry of 1977 and during the commissioning of THORP in 1993 and 1994.' The report of February 2000 (*Storage of liquid high level waste*) is an important document for anyone concerned about the presence of dangerous high-level waste at Sellafield. Its authors concluded that BNFL's plans (i.e. its 'safety case') were 'adequate to support current and future operations'. The inspectorate made a number of specific recommendations to ensure that operations will continue to meet the government's requirement 'to control and reduce risks ALARP [as low as is reasonably possible]'. However, the authors of the report never anticipated an attack on a nuclear power plant similar in nature to that on the Twin Towers of Manhattan.

Stay Cool

Some fears about Sellafield would be allayed if the highly active liquids stored in B215 were quickly turned into a solid glass form and removed to a long-term underground storage facility. However, the vitrification of Sellafield's accumulated waste has proceeded neither as quickly nor as well as was first envisaged by BNFL. The three vitrification lines are reported to have operated at only 35 per cent of their intended production capacity, because of certain technical difficulties. Moreover, there is no British long-term storage facility for spent nuclear fuel. Yet, as the HSE put it in 2000, 'the storage of solid High Level Waste (HLW) in the VPS [Vitrified Product Store] has advantages over storage as a liquid in the HASTs [Highly Active Storage Tanks], because the fission products are immobilised in the solid ['glass'] matrix, and the glass is cooled by the natural circulation of air'. Such cooling is an essential part of keeping radioactive waste safe. The cooling which is achieved by the natural circulation of air, unlike cooling in other areas of waste storage, does not depend on the continued availability of installed services such as electricity and water, which can fail due to human error or sabotage or technical defect. It is sometimes described as being a 'passively safe' technique and is, clearly, superior to the form of cooling on which the safety of storage tanks full of dangerous liquid now depends.

Intermediate-Level Waste

Intermediate-level waste (ILW) is said to be 'far less radioactive' than high-level waste and includes materials such as fuel element cladding, contaminated equipment and certain sludges resulting from various treatment processes. It can be stored in concrete silos, or mixed into cement blocks inside steel drums for eventual storage. Intermediate-level waste arises mainly from the dismantling of spent fuel, and from the general operation and maintenance of the radioactive plant. It constitutes about 6 per cent of all radioactive waste by volume and is currently stored mainly at the sites of production. The term 'transuranic' is sometimes used to refer to certain intermediate-level waste.

Between 1949 and 1982, 73,530 [metric] tonnes of low-level and intermediate-level waste were loaded into drums and dumped by the United Kingdom in the North-East Atlantic Ocean. Since 1982, intermediate-level waste which would have been disposed of at sea has been stockpiled. In addition, some 'arisings' from the late 1940s onwards have been stored on particular sites. In 1993, the United Kingdom government accepted an international ban on sea disposal of radioactive wastes. But its officials did not fish out what was already down there.

Low-Level Waste
Low-level waste (LLW) includes items such as paper towels, clothing and laboratory equipment which have been used where radioactive materials are handled. It may also refer to certain parts of decommissioned power plants. It also includes low-level liquid waste from cooling pond water.

Liquid 'low-level' waste from Sellafield has been released into the Irish Sea as a matter of course, after treatment of the waste to reduce its radioactive content. Over the years, BNFL has taken steps to limit its radioactive discharges into the Irish Sea. The company says that it has invested 'hundreds of millions of pounds to treat this "slightly contaminated water" and to reduce liquid discharges'. BNFL assures the public that the radioactive content of its low-level waste is 'very small'. However, J. Samuel Walker, in his recent book *Permissable dose: a history of radiation protection*, considers the absence of definitive evidence on the effects of 'low-level' exposure to radioactivity.

The UK's Department of Trade and Industry says that,

> Low-level liquid wastes arise from water used in cooling, cleaning and other operational processes. Low-level gaseous wastes arise from nuclear plant operations and ventilation systems. Both are discharged to the environment, after treatment to reduce their radioactive content, under authorisations granted under the Radioactive Substances Act 1993.

The Department of Trade and Industry further states that,

> Improvements in technology have allowed levels of radioactivity in liquid and gaseous discharges to be progressively reduced over the last 20 years. For example, as a result of investing over £750 million in new treatment plants at Sellafield, BNFL's radioactive discharges to the Irish Sea are now less than 1 per cent of what they were in the mid-1970s.

Drigg

Since 1959, about one million cubic metres of 'low-level' radioactive waste (LLW) have been disposed of in Britain, some of it coming from the Republic of Ireland. Solid 'low-level' objects, including equipment from the energy and medical sectors such as gloves, overalls or laboratory equipment, are placed in earthen trenches at a 300-acre purpose-built disposal facility at Drigg, a few miles south of Sellafield. According to the UK Department of Trade and Industry, solid 'low-level' waste constitutes about 94 per cent by volume of all radioactive waste. Solid waste has been dumped, to a lesser extent, at Dounreay, Caithness, in the far north of Scotland. Low-level waste scheduled for Drigg is now usually subjected to high-force compaction before being placed in metal containers of about 15 cubic metres capacity, prior to grouting with cement and placement inside a concrete-lined vault. BNFL operates the Drigg site as a commercial venture.

In addition to the low-level waste generated by BNFL, which is said to constitute about three-quarters of what goes to Drigg, the facility is available as a disposal service to a spectrum of customers throughout the United Kingdom and beyond, including hospitals and universities. In its annual report for 2001, BNFL described Drigg as 'the only route for the disposal of this low-level waste' and as 'a national asset'. In recent years, increasingly costly techniques have been used to compact waste at the site to maximise the remaining capacity and to delay, for as long as possible, the point at which the existing site becomes full. According to the 2002 British White Paper on managing the nuclear legacy, Drigg is expected to remain open until 2060.

Nirex

In the early 1980s, the nuclear industry, with the agreement of the United Kingdom government, established a company called Nirex to examine safe, environmental and economic aspects of deep geological disposal of radioactive waste. More than 95 per cent of all of the United Kingdom's radioactive waste is said to come from the nuclear power industry. Based at Harwell in Oxfordshire, Nirex deals with intermediate-level waste, which accounts for the majority of radioactive waste currently in temporary storage, and also deals with some low-level waste. From 1987, Nirex began a process of searching for an underground site to serve as a long-term nuclear dump (or 'repository' as such facilities are also known). About 500 possible sites are said to have been identified, but a location at Sellafield was actually chosen. As noted earlier, there is already a dump for low-level waste at Drigg, a short distance from Sellafield.

Down Under the Farm

The location selected by Nirex to become a dump was Longlands Farm, owned by BNFL and lying south-east of the existing Sellafield complex. The choice of this proposed site was hardly surprising, because most of the waste to be disposed of by Nirex comes from activities at Sellafield itself. Nirex wanted to bury 400,000 cubic metres of radioactive waste under Longlands Farm.

Nirex accepted that, in order to calculate the risks to human health and society from such a repository, a 'scientific understanding of the behaviour of the facility, its environs and its radioactive contents over a time-scale of many thousands of years' was required. This is quite a tall order. Nirex was, more specifically, required by official safety requirements to establish that the risk of someone contracting fatal cancer because of the release of radioactivity from the repository would be less than one in a million each year.

In 1994, Nirex applied to Cumbria County Council for planning permission to develop an underground 'Rock Characterisation Facility (RCF)' beneath Longlands Farm. This facility was to have taken the form of a subterranean laboratory, which would have allowed further investigations to confirm whether or not the site was suitable for a dump. Cumbria County Council rejected the application, which was then appealed by Nirex. As part of the appeal process, a lengthy Planning Inquiry began in September 1995, and ended on 1 February 1996. Objections to any possible permission for a storage facility under Longlands Farm were made by a wide grouping from British universities and consulting companies, co-ordinated by Cumbria County Council and Friends of the Earth and Greenpeace. These objections were subsequently published by the University of Glasgow.

In March 1997, Nirex lost its appeal. Its proposal for a dump under Longlands Farm had been rejected. Nirex responded to the defeat by reverting from specific design development for Longlands Farm to the development of a generic repository design. This 'fits all reasonable sizes' design could be applied to other potential sites in general. On the basis of its ongoing research and design, Nirex has continued to advise the producers of nuclear waste on methods of packaging their radioactive waste such that they will conform to the requirements that Nirex believes will be appropriate for their safe disposal in any future underground repository.

'Legitimate' Irish Interest

The government of Ireland had made known its strong objections to the plans for a dump at Longlands Farm. Subsequently, it took note of the fact that the then United Kingdom Secretary of State, John Gummer, had concurred with certain conclusions of the Planning Inspector regarding the concerns of the Irish government. Gummer agreed, in the words of his officials, that 'the people of Ireland have a legitimate interest in any proposal for a repository for radioactive waste near the Irish Sea coast. He is acutely aware of the UK government's obligations to other states which are set out in various international obligations in respect of the sea and the environment more generally.'

Nirex: Still Planning Dumps

Following the failure of its application for Longlands Farm, Nirex issued a frank acknowledgement of its own shortcomings. It accepted that it had made mistakes. Referring to the rejection of its application, Nirex states today that, 'Although much of Nirex's scientific research was praised, Nirex was criticised for the way it had carried out its work in other areas, particularly in selecting the site. We at Nirex have looked very hard at ourselves following the failure of the programme, and we have tried to address the criticisms that we were a secretive and arrogant organisation.'

Nirex claims to be working on an intensive internal programme to develop and implement a transparency policy. This policy, says Nirex, 'covers key areas such as making available the information we hold and having constructive discussions with all stakeholders, that is, the individuals or groups who have an interest in what we are doing'. Meanwhile, radioactive waste has continued to accumulate at various locations that are almost certainly less safe than would be an underground dump designed by Nirex to the highest possible standards.

At the present time, Nirex has several roles, which it defines as follows:

- Advising the organisations and companies that produce radioactive waste on how they should package the waste.
- Setting the standards for radioactive waste packaging. Nirex monitors the processes of these organisations and companies to check, for example, that they have procedures in place to ensure that adequate records are kept.
- Producing on behalf of the Department of the Environment, Food and Rural Affairs an updated public record of the quantities and types of radioactive waste that exist currently in the UK.
- Continuing to develop an understanding of the options for dealing with radioactive waste. Nirex discusses with a wide range of organisations the many common issues that we all face. These can be scientific, technological and environmental. This also includes developing an understanding of the requirements for public acceptability. Nirex programmes help to keep the UK abreast of international expertise in research and development into the disposal of radioactive waste.

The government of the United Kingdom is currently reviewing future options on what to do with radioactive waste. In the meantime, Nirex says that it is rethinking its own role. Further information on Nirex is available on its website at http://www.nirex.co.uk/iabout.htm.

A New Authority

The new UK Liabilities Management Authority will take on certain responsibilities in relation to nuclear waste of all types that is stored 'short-term' on the many sites which it will control. The role of the LMA was considered already. It is not, at the time of writing, intended to take over the role of Nirex in relation to the long-term storage of waste.

Yucca Mountain

Up to one thousand tests of nuclear weapons have been conducted in Nevada, in the United States. It is now proposed to construct in that same state the first American long-term nuclear waste repository. The location selected is Yucca Mountain, about one hundred miles north-west of Las Vegas. Many Americans, including the state authorities of Nevada, oppose the proposed dump.

Harry Henderson, in his recent book *Nuclear Power*, (p. 90), makes the point that the desert climate of Nevada 'is important because water movement is the primary means by which radioactive waste could be transported from a repository'. Reading this, one wonders about the choice by Nirex of a place such as Longlands Farm in wet Cumbria, by the sea, for its proposed dump. The United States government claims that the deep, underground facility at Yucca Mountain will be a safe store for radioactive waste for at least 10,000 years. Critics contest its claim, and also query why 10,000 years is considered adequate. A detailed and fascinating report on the Yucca Mountain proposal appeared in *National Geographic* magazine in July 2002 ('Half-life: the lethal legacy of America's nuclear waste', pp. 2–33). According to its authors, 'the detritus of the nuclear age includes 52,000 tons of spent fuel from commercial, military, and research reactors, as well as 91 million gallons of radioactive waste from plutonium processing'. There remains an inescapable need to create as soon as is practical the safest possible long-term storage facilities for existing stockpiles of nuclear waste.

Waste Won't Go Away

Even in the event of the British nuclear industry closing down completely tomorrow, an enormous volume of radioactive waste would remain to be dealt with. If not properly looked after, this waste could cause serious harm to humans and to the natural environment. Nirex defines the problem as 'an ethical issue that our society needs to address'. According to Nirex,

> People might have different views on nuclear power or weapons, but the fact remains that for over forty years nuclear power has been providing electricity in Britain and nuclear weapons have been retained as a matter of policy by successive governments. It is now the job of this generation to deal with the resulting wastes. We believe we cannot put off the problem for future generations to deal with. We need to decide what to do with the long-term radioactive waste that already exists.

It is not only the British who must face up to the problem of nuclear waste. That is why, on 6 November 2002, the European Commission put forward a draft directive which envisages a common and binding framework for the disposal of nuclear waste by any Member State. However, no over-arching corps of European inspectors will be set up, and national agencies will continue to have the prime responsibility for inspections. The Commission acknowledges that 'nuclear safety cannot be guaranteed unless sufficient financial resources are set aside' and says that 'the present funding allocated to research on waste management is insufficient'.

The Commission is committed to geological burial of waste 'as the safest method of disposal known at present'. If its proposals are accepted by EU governments, then Member States will have to adopt, according to a pre-set timetable, national programmes for the disposal of radioactive wastes, including, in particular, deep burial of highly radioactive wastes. And they will have to decide on the location of such burial sites (whether national or shared by several States) for highly radioactive wastes by 2008 at the latest and to have the sites operational at the latest by 2018. For low-activity, short-life waste, the necessary disposal arrangements must be ready at the latest by 2013.

The text of the Commission's proposals and other relevant information may be found at http://europa.eu.int/comm/energy/nuclear/index_en.html.

Return of the (Radioactive) Mummy
What if Nirex succeeds in getting permission for a long-term dump of highly radioactive waste under some British farm or mountain, as it surely must eventually? You can wrap it. You can encase it. You can bury it in concrete beneath the earth. But can you contain it for thousands and thousands of years? The disposal of nuclear waste is a growing problem, as additional waste is produced by the nuclear industry every day. Yet, there are very many events, features and processes that may affect the rate at which radioactivity from a nuclear dump is released into the environment. The dangerous factors include meteorite impact and solar isolation, earthquakes, an ice age, plant and animal evolution, human error, sabotage, river meander and urban development. What if, in a time of terrible war or plague, no State could sustain the expertise and expenditure essential to keep safe a dump such as that proposed for Yucca Mountain?

Nirex points out that the first engineered barrier against any escape of radioactivity from a long-term nuclear repository would be the waste package itself, comprising the container into which the waste is packaged, together with the matrix (typically cement) that holds the waste in place within the container. According to Nirex, research indicates that this first barrier will provide a high level of containment for at least one thousand years, by which time (it is said) about 99 per cent of the 'activity' (radioactivity) will have decayed.

In a Nirex repository, the waste packages would be surrounded by cement-based material and provide a chemical barrier to the movement of radionuclides. This 'should' remain effective for at least one million years, time enough (it is said) for 99 per cent of the remaining 1 per cent of the activity to decay. The final barrier, says Nirex, is the natural one, the stable rock formation at a depth to isolate the wastes from the effects of natural processes such as ice ages, and also from human intrusion. It accepts that the host rock must also provide groundwater conditions that enable the repository to perform effectively.

Future Generations
By burying radioactive wastes now, we are creating a challenge and a danger for future generations. The kinds of delight experienced by people who discover the intricacies of a ceremonial complex such as Newgrange, that ancient Irish burial site in the valley of the River Boyne, will never be experienced by those who stumble upon a nuclear store in thousands of years to come. What if wars or plagues or natural disasters make such repositories unviable or unstable?

What if the passage of time and unforeseen eventualities result, somehow, in the location of such dumps being forgotten? It was such an idea, sparked by reading about an existing underground store for 'transuranic' nuclear waste in the south-western United States, that inspired me to write the following poem, included in a collection of my poetry published by Kestrel Books in 1995 (*Standing on Bray Head; hoping it might be so*).

Today, for 'New Mexico' one might read 'Nevada', because of the recent designation of Yucca Mountain as a planned dump. And it must be admitted that the Native Americans of at least one 'reservation' have welcomed the storage of nuclear waste as a means of generating local revenue. On certain other reservations, they have preferred to build gambling casinos, taking advantage of a loophole in the gaming laws of the United States. Casinos constitute one form of betting on the future. Nuclear waste dumps constitute another.

Newgrange to New Mexico

Try then to bury our lost innocence?
Out under the desert of New Mexico,
in the eyes of the Native Americans,
the genocide piled high on genocide.

Try then to advise them not to pry?
Who maybe living millennia at peace
must understand what we now say,
for all time, – they've been warned!

Try then to fix signs they can read?
We can't undo the dirty deed done:
the box is open and the clocks run,
generations draw breath and wait.

What did they try to teach us then?
Near the banks of the River Boyne,
who, over four thousand years ago,
erected mounds to honour the sun.

Those circles and spirals and loops?
The big triangles and strange hoops:
symbols and signs of consciousness,
speaking in riddles down the years.

Long before Egypt had its pyramids,
builders at Knowth and Newgrange
aligned a burial-chamber on a star,
respectful then of the environment.

If, at the winter solstice every year,
December skies are clear at sunrise,
then rays creep in along its passage
lighting the mound's inner sanctum.

Just a few short thousand years ago,
they too decorated their handiwork
with simple marks or basic designs:
Newgrange stands for natural unity.

So what has New Mexico got to say?
What, in ten thousand years hence?
Will future people, stumbling on it,
enter its passages to seek meaning?

There, on the walls they'll see signs,
created by today's genius scientists:
will they pause and decipher these
or, like at Newgrange, just walk on?

As at Newgrange, they'll find death,
but not as old bones turned to dust:
in New Mexico the sands are quick,
where we put nuclear waste on ice.

Try then to say we knew not what?
That people once lived with nature,
courting the sun's energy with life?
That we played God to future ages?

CHAPTER 16

British Standards

The people of Ireland are largely dependant upon the authorities of the United Kingdom when it comes to ensuring that Sellafield is being run safely. While there is also some degree of international accountability, the United Kingdom itself is responsible for regulating the British nuclear industry in accordance with acceptable standards. These standards are set in the context of international agreements relating to nuclear energy.

The principal organisations which regulate and police the British nuclear industry are the Health and Safety Executive (through the Nuclear Safety Directorate and its NII), the environment agencies and the OCNS. Government departments also have particular responsibilities. Before examining the role of the regulators, which is necessary in order to understand how Sellafield is licensed and monitored, it is worth considering some of the concepts that lie behind the policies and decisions of those regulators.

B, BEP, ALARAP...
Can you tell the difference between ALARA and ALARP? While neither formula will wash away those nasty radioactive stains, both are intended to prevent spills in the first instance. And you can always rely on 'BPM' or on 'BEP'.

Government departments in the United Kingdom have agreed a formula of care for the operators of nuclear installations. These official 'principles for the protection of the human environment' are intended to ensure that exposures from radiation are 'as low as reasonably achievable', taking into account economic and social factors.

This 'ALARA' principle is founded on the Euratom Basic Safety Standards Directive 96/29, which has been transposed into the law of the United Kingdom. 'Essentially', state the authors of the recent British White Paper on liabilities management, 'it means that all reasonable steps should be taken to protect people. In making this judgement, factors such as the costs involved in taking protection measures are weighed against benefits obtained, including the reduction in risks to people.'

The official principles adopted by the United Kingdom intone solemnly that,

> During the institutional management and post institutional management periods, the authorising departments will require operators to use the best practicable means to ensure that any radioactivity coming from a facility is as low as reasonably practicable (ALARP). The effect of this will be to ensure that exposures are ALARA.

And that's not all:

> The authorising departments will also, in conjunction with the waste producers, apply the ALARA principle to decisions between overall management procedures relating to specific wastes and covering treatment of the waste, method of immobilisation, and method of disposal.

Which is quite a lot of ALARA.

The United Kingdom's Health and Safety Commission points out that,

> no human activity is entirely free from features that are detrimental or involve risks to life, and it is never

possible to be sure that every eventuality has been covered by safety precautions. The basic policy of the Health and Safety Commission is to eliminate these ill effects so far as is reasonably practicable . . . taking account of known technology and costs.

The policy has been put another way by the Nuclear Installations Inspectorate of the United Kingdom:

This means that the legal duty is to reduce risk unless the cost of these measures (whether in money, time or trouble) can be shown to be disproportionate to the risk that would be avoided. In short, risk must be reduced to a level which is as low as reasonably practicable which is known as the ALARP principle.

So we should all be safe, so long as it does not cost too much. Besides, as well as ALARA and ALARP protecting us, we may also rely on the British nuclear industry abiding by 'BPM' (Best Practical Means). This is a term used in authorisations issued under the United Kingdom Radioactive Substances Act. It is understood to give effect to the consensus of contracting parties to the OSPAR Convention, who are said to be committed to the application of 'the precautionary principle' and 'the polluter-pays principle', and to the application of 'Best Available Techniques (BAT)' and 'Best Environmental Practice (BEP)', including, where appropriate, 'Clean Technology'. Well, thank goodness for that. The British point out that the 'BPM' principle

requires operators to take all reasonably practicable measures in the design and operational management of their facilities to minimise discharges and disposals of radioactive waste, so as to achieve a high standard of protection for the public and the environment. BPM is applied to such aspects as minimising waste creation, abating discharges, and monitoring plant, discharges and the environment.

The application of 'BPM' is said to take account of such factors as the availability and cost of relevant measures, operator safety and the benefits of reduced discharges and disposals. According to the recent British White Paper on liabilities management, 'If the operator is using BPM, radiation risks to the public and the environment will be ALARA.'

So we should all be safe, so long as it remains practicable and affordable for BNFL to protect us.

Buck Stops There

BNFL itself is ultimately responsible for the safety of its various operations at Sellafield, including reprocessing. Certain official regulations and laws of the United Kingdom are intended to ensure that the nuclear power industry is run to an acceptable minimum standard of safety. However, BNFL does not exhaust its duty of care to the public by simply complying with particular regulations.

In 1981, the NII made it clear that when it comes to safety the buck stops on the desk of BNFL and not on that of the NII. A report from the NII noted that,

> We must also emphasise that the responsibility for establishing adequate arrangements for health and safety, and for a system of audit to ensure that it is working effectively in practice, rests with the management of the Company [BNFL]. They are not to rely for this on the Nuclear Installations Inspectorate of the [Health and Safety] Executive. Our responsibility is providing an independent check that the Company is fulfilling its own responsibilities. (cited in Charlesworth *et al.*, *Windscale: the management of safety*, p. 5).

Safety Cases

The regulatory regime for the United Kingdom's nuclear industry is essentially non-prescriptive. It is up to BNFL, and other relevant bodies, to demonstrate to the NII that safety is properly controlled on their sites and that the conditions of their licences are being met. An important part of the process is the presentation of periodic 'safety cases'. The term 'safety case' is used to describe one or more documents that the operator of a nuclear power plant, or a proposed nuclear power plant, must produce to satisfy the NII that a particular licence is to be issued. They include a preliminary safety report, pre-construction safety report and pre-operational safety report. The 'safety case' documents describe the principal features of an item or system or plant, how it is to be managed, designed, installed, tested, maintained, operated, taken out of service, or decommissioned so that nuclear safety will be ensured. Agencies responsible for monitoring the safety of nuclear installations refer back to the safety cases of particular facilities. The terms 'safety report' and 'safety submission' are sometimes used instead of 'safety case'.

Official Regulation

A number of government departments and other organisations have a role to play in monitoring and enforcing safety standards at various British nuclear sites, or in respect to the transportation of radioactive materials between them. The four departments most directly responsible are (i) Trade and Industry (DTI), (ii) Transport, (iii) Environment, Food and Rural Affairs, and (iv) Defence.

DTI

In England and Wales the Secretary of State for Trade and Industry is responsible for nuclear safety policy, and ultimately oversees the licensing regime for nuclear facilities.

In addition to its regulatory activities, the DTI is also concerned with ensuring that the views of the nuclear and certain other industries are represented in the decision-making process determining radioactive waste management policy. The DTI further co-ordinates policy at national level as lead government department when it comes to the arrangements for responding to accidents or other emergencies with off-site effects from a licensed civil nuclear site in England and Wales.

For more on its role and on the nuclear industry see http://www.dti.gov.uk/nid/index.htm.

OCNS

The Office for Civil Nuclear Security operates as an independent unit within the Department of Trade and Industry, with full autonomy in regulatory and operational matters. A number of references have already been made to its work. During 2002, as was noted earlier, its director published a major report on its responsibilities and operations, which are of particular interest in the aftermath of the attacks of 11 September 2001.

Health and Safety Commission

The Health and Safety Commission (HSC), which advises the Secretary of State for Trade and Industry, is obliged by law to ensure that there is an adequate framework for the regulation of safety at British nuclear sites. This commission was established under the Health and Safety at Work Act 1974. Its licensing and day-to-day regulation is carried out by the Health and Safety Executive (HSE), which regulates work-related health and safety generally. Thus, both the Health and Safety Commission and its Health and Safety Executive are ultimately accountable to the Secretary of State for Trade and Industry for their nuclear safety work.

NuSAC

The Secretary of State for Trade and Industry and the Health and Safety Commission are, in turn, advised on policy matters by the Nuclear Safety Advisory Committee (NuSAC). This is an independent body that replaced the former Advisory Committee on the Safety of Nuclear Installations.

Health and Safety Executive

The Health and Safety Executive of the United Kingdom operates, as has been said, under the direction of the Health and Safety Commission. It is the executive arm of the HSC.

The Health and Safety Executive is also the official body that grants licences for the operations of nuclear power plants. The conditions attached to the licences that it issues cover all the arrangements for managing safety, including response to accidents, leaks and spillages of radioactive materials, emergency planning arrangements, and all aspects of the transportation of radioactive material on sites.

The licensing function of the HSE is administered on its behalf by its Nuclear Safety Directorate (NSD) and its Nuclear Installations Inspectorate (NII). These bodies are responsible for the safety rather than the security of nuclear sites. Security, as we have seen, is a matter for the OCNS to supervise. A small unit of the HSE, known as Safety Policy Directorate E, is separate from the Nuclear Safety Directorate but helps to develop policies in relation to nuclear safety.

The policy of the Health and Safety Commission is to eliminate ill effects from human activities 'so far as is reasonably practicable'. This, as was seen above, is the 'ALARP' principle. In relation to the normal operation of nuclear installations, this basic policy is expressed in the form of a requirement that nuclear operations shall not give rise to levels of exposure of workers and members of the public to radiation and radioactive materials that exceed certain limits. These limits were recommended by the International Commission on Radiological Protection and incorporated into a Euratom Directive on radiation protection standards. There is a further, additional requirement that is also part of the international recommendations and the Euratom Directive. It stipulates that any exposure to radiation should be reduced as far below these limits as can be reasonably achieved, taking costs into account in judging what is reasonable. This is the formula known as the ALARA principle.

Nuclear Safety Directorate
Under the United Kingdom's Nuclear Installations Act 1965 (NIA), a site cannot have a nuclear plant on it unless the user has been granted a site licence. As mentioned above, this

licensing system is administered by the HSE's Nuclear Safety Directorate and its NII. NSD (formerly known as the Nuclear Safety *Division* rather than *Directorate*) consists of four divisions which deal with, respectively, British Energy plc; BNFL; Nuclear Safety Research and Strategy; the United Kingdom's Atomic Energy Authority, Defence and other sites. Thus, BNFL holds its current licence for Sellafield from the Nuclear Safety Directorate. NSD, acting for HSE, attaches conditions to this and to other site licences relating to the general safety requirements for preventing or dealing with the risks of a nuclear accident. Licensees, including BNFL, must comply with these conditions in various ways. They will, for example, prepare what is known as 'a safety case' (see above) to meet relevant needs at a particular stage in a plant's life, or agree arrangements to meet a licence condition.

Guidance for the nuclear industry is also set out in the safety assessment principles, which NSD has developed for its own use and made available to the public. NSD seeks to keep up and improve safety standards for work with ionising radiation at licensed nuclear installations. It does so through its licensing powers by assessing safety cases and inspecting sites for licence compliance. It sets national regulatory standards and helps to develop international nuclear safety standards.

Most of NSD's technical staff were employed in industry for ten years or more before joining NSD, usually in the nuclear industry. All of the technical skills found in any engineering-based industry are said to be found in NSD, as well as reactor physics and radiological protection. Specialised areas of expertise are needed, such as pressure vessel technology, metallurgy, seismology and heat transfer, and there is said to be a strong emphasis on human factors, quality management and the management of safety. NSD also uses consultants, and has a nuclear safety studies programme geared to its own needs. These give NSD an independent source of specialist advice. Its consultants come from universities, engineering firms and national organisations, such as the Welding Institute and the British Geological Survey.

Every year, the Nuclear Safety Directorate produces an *Annual Research Index*. This lists the issues that NSD believes need more research and will form the basis of the research procured under the Annual HSC Co-ordinated Nuclear Research Programme. NSD gives specialist assistance to the International Atomic Energy Agency (IAEA), to the Organisation for Economic Co-operation and Development's Nuclear Energy Agency and to Euratom. The NSD, and the Health and Safety Executive itself, also share information with nuclear regulators in other countries, including Ireland.

Nuclear Installations Inspectorate (NII)
Located within the Nuclear Safety Directorate of the Health and Safety Executive is the Nuclear Installations Inspectorate (NII). The inspectorate was originally established in 1960, under powers in the Nuclear Installations (Licensing and Insurance Act) 1959, and thus predates the Health and Safety Executive itself. It was set up as a direct result of the fire at Windscale/Sellafield in 1957. The inspectorate is concerned with the safety of operations, the prevention of accidents and the protection of the workforce and the public. It is not responsible for security matters. In practice, it has a very high workload.

In 1971, for the first time, the NII was given responsibilities for supervising Windscale/Sellafield. It found that 'the standard of the plants there had deteriorated to an unsatisfactory level' and set about ensuring that arrangements were made by BNFL for achieving a high standard of safety. Some essential plant improvements required substantial design and development work.

The NII used to report directly to the Secretary of State for Trade and Industry. In 1975, it became part of the Health and Safety Executive (HSE), when the HSE was formed under the Health and Safety at Work Act 1974. There are three principle functions of the inspectorate:

• 	Assessing the safety aspects of new nuclear systems that may be introduced into Britain. This includes new types of reactors and new types of chemical plants for making, treating and reprocessing fuel.

- Assessing the safety of specific designs of reactors, or other plants, for which a licence is being sought. This assessment results in a licence being issued, with conditions.
- Inspecting nuclear plants, to ensure that the conditions in licences are being met and to assess any changes in operations.

As stated earlier (p. 193), the NII has made it clear that the ultimate responsibility for safety rests with BNFL and not with the NII. More recently, the NII has pointed out that the United Kingdom's regulatory regime is 'essentially non-prescriptive'. In its own conditions attached to licences, the NII sets goals but does not specify the means of achieving them. Each licence applicant must propose their own solutions to safety issues and, as we saw, provide a 'safety case' to demonstrate that safety will be properly controlled through all stages of a plant's life.

The NII has an annual liaison meeting with the Radiological Protection Institute of Ireland. *The work of HSE's Nuclear Installations Inspectorate*, a useful booklet, is available from the Nuclear Safety Directorate of the Health and Safety Executive. Although printed in 1995, it is still substantially accurate on the job done by the NII.

The Health and Safety Executive's *Nuclear safety newsletter* is published every February, June and October. A *Statement of nuclear incidents at nuclear installations* is also published quarterly by the HSE. These documents and further information about the Nuclear Safety Directorate and the NII are available at http://www.hse.gov.uk/nsd/nsdhome.htm.

Transport
The Secretary of State for Transport accounts to parliament for the safe transport of nuclear materials within Britain. The relevant regulators are the same as those for transport safety

generally; for road, the Department for Transport itself; for rail, the Health and Safety Executive; for sea, the Maritime and Coastguard Agency; for air, the Civil Aviation Authority.

Environment
The job of regulating the routine discharge and disposal of nuclear waste and other radioactive materials is that of the Environment Agency in England and Wales, and of the Scottish Environment Protection Agency in Scotland. The Environment Agency is accountable to the Secretary of State for Environment, Food and Rural Affairs for its work in England, and to the National Assembly for Wales for its work in Wales. The Scottish Environment Protection Agency is accountable to the Scottish Executive. These two environmental agencies authorise and control radioactive discharges and waste disposals to air, water and land. They also regulate the storage and use of radioactive substances and the accumulation of radioactive waste, except on licensed and defence-related nuclear sites where these functions are a key part of day-to-day operations and need to be regulated in that context. On such licensed nuclear sites, the storage and use of radioactive materials and the accumulation of radioactive waste is (as we have seen) regulated by the Health and Safety Executive. The web address of the English and Welsh Environment Agency is http://www.environment-agency.gov.uk/ and that of the Scottish Agency is http://www.sepa.org.uk/.

Defence
The Secretary of State for Defence is responsible for the nuclear weapons and naval nuclear propulsion programmes, and so accounts to parliament for safety at all defence-related nuclear sites. Some of these sites are licensed by the Health and Safety Executive in the same way as civil sites, but others are not, because 'the Crown' (the UK state) is exempt from some of the relevant legislation. The licensed sites are regulated in the same way as civil licensed sites, except that

the Health and Safety Commission and Executive are accountable to the Secretary of State for Defence for this work. On unlicensed defence-related nuclear sites, the Secretary of State for Defence operates a policy of voluntary compliance with the regulatory regime operated by the Environment Agency and Scottish Environment Protection Agency. It is claimed that 'standards are at least as good as those required by civil regulation where reasonably practical'. Moreover, 'where appropriate', internal arrangements are overseen by the Health and Safety Executive.

United Kingdom Atomic Energy Authority (UKAEA)
The UKAEA was founded in 1954 and pioneered the development of nuclear energy in Britain. A statutory body, it became established as one of Britain's largest and best-known research institutions, employing thousands of people. The UKAEA quickly attracted massive government funding for reactor research and development. Out of part of its structure, in 1971, came BNFL. As noted earlier, BNFL was created to operate the fuel enrichment, fuel manufacturing and fuel reprocessing services developed by the UKAEA and to run the Calder Hall and Chapelcross reactors. The UKAEA subsequently spent many years and much money pursuing the dream of an efficient fast breeder nuclear reactor, among other research projects.

In 1988, Michael Flood undertook for Friends of the Earth a consultancy report on the performance and future of the UKAEA. That report, *The end of the nuclear dream: the UKAEA and its role in nuclear research and development*, was highly critical of the authority, claiming that a series of errors of judgement about reactor design and research priorities had left the United Kingdom's nuclear industry in a weak position.

Today, the UKAEA is responsible for managing the decommissioning of those nuclear reactors and other radioactive facilities that have been used for the United Kingdom's nuclear research and development programme. It is required to do so in a safe and environmentally sensitive manner. Its objective is, essentially, to restore such sites for conventional use. This involves the decommissioning of redundant facilities, managing the resulting wastes in a safe and environmentally responsible way, progressively delicensing decontaminated sites and promoting, when possible, alternative uses for sites. The UKAEA says that its top priorities are safety and the protection of the environment, 'while delivering value for money to the taxpayer'. In developing nuclear technologies on behalf of the UK government it is committed to working with industry to implement safe, innovative technical solutions. Insofar as the UKAEA still has a role in relation to some facilities at the Sellafield site, it works in conjunction with BNFL when appropriate. More information on it may be found at http://www.ukaea.org.uk/about/index.htm.

Liabilities Management Authority
As noted earlier, a special Liabilities Management Authority (LMA) is intended to take over the duty of caring for Britain's old nuclear infrastructure, on behalf of the state. The LMA's proposed brief is said to represent about 85 per cent of total nuclear liabilities in the United Kingdom and is wholly the responsibility of government.

Decommissioning such infrastructure will be no easy task. Today, nuclear plants are built with a view to their being dismantled eventually. It was not always so. As the authors of the 2002 British White Paper on the LMA point out, 'many legacy facilities were built and used at a time when regulatory requirements and operational priorities were very different to those that apply today. Early operating records and waste inventories are often incomplete.' The authors remark that the uncertainties that flow from this situation are fundamental to understanding the problems involved in legacy management and clean up. In some instances, they say, 'the biggest challenge is not deciding how to tackle a particular task but working out what exactly has to be dealt with. Equally important, many facilities are one-offs, built to test the feasibility of, or prove the commercial viability of, different technologies. There are no simple problems and few simple solutions.' The White Paper on the Liabilities Management Authority is essential reading for anyone interested in the future of the nuclear energy industry and its wastes.

Irish Concerns

T he government of Ireland says that it 'has firmly rejected nuclear energy'. Within the Republic of Ireland, there is currently a broad political consensus on the issue of nuclear power. The Irish government's policy places a heavy emphasis on nuclear safety, radiological protection and emergency preparedness.

Not Always So
It was not always so. During the 1970s, the Irish government had its own plans to build a nuclear power plant at Carnsore Point in County Wexford. Those plans faced intense opposition, but Ray Burke, who was then a junior minister at the Department of Industry, Commerce and Energy, decreed loftily that, 'a public inquiry in relation to a technology which is already being successfully applied in many countries would hardly be a helpful exercise' (cited in Carroll and Kelly, *A nuclear Ireland*, p. 38).' Nevertheless, its plans for a nuclear power station were later dropped by the government. Had the Irish then proceeded, a nuclear plant at Carnsore would probably have been similar in type to the Magnox station at Wylfa, on the island of Anglesey in Wales. There, two reactors commenced operations in 1971. Wylfa is BNFL's largest power station. It receives little publicity, although its existence constitutes at least as great a danger to Ireland as that of any other power station in Britain.

Instead of opting for nuclear power, the Irish invested heavily in power stations that are dependant on fossil fuels. These now belch out a significant proportion of Ireland's greenhouse gases. There is no nuclear power plant in Northern Ireland, partly for reasons of local security.

The Irish and Sellafield

For years, the Sellafield complex of BNFL has been a source of grave concern to successive Irish governments and to the Irish public. Ask an Irish person how near is Sellafield to the shores of Ireland and he or she is likely to answer 'too near'. Approximately at the same latitude as Belfast, the Cumbrian nuclear site is just 112 miles from the Irish coast at its closest point (at Clogher Head). Sellafield faces west to the Isle of Man and Ireland.

One historian of the Sellafield area, Marjorie Rowling, has written that,

> Since the sea, especially in the past, was a highway rather than a barrier, these localities have had a close cultural link with our district. On the stones of the Cumbrian prehistoric circles are found carvings of cups, concentric rings with radical channels, spirals and other motifs which are found in the Neolithic passage graves of the Boyne in Ireland, as well as on stones in Scotland and Northumberland.

Perhaps it is some deeply rooted social memory of the practical proximity of Sellafield that makes Irish people so adverse to the storage of highly radioactive waste, and to the reprocessing of plutonium, on the site.

The direction of winds in Cumbria is not constant, and local breezes sometimes blow towards Ireland. Not suprisingly, Irish people fear that the devil's brew of spent fuel and other concoctions at Sellafield is contaminating both the air and sea, creating a permanent and potentially deadly link between these two islands. They hope that Sellafield meets the fate of another ambitious engineering project that is said to have failed in the time before records were kept. For tradition has it that the Devil decided to link the Isle of Man to Cumberland by a bridge starting from the Herdy Neb, a promontory near Seascale. 'But', writes Marjorie Rowling, 'as he was carrying the foundation stone, his apron strings broke and the boulder fell. This still bears two white strings – as can be seen on Carl Crag, the name given to the stone the Devil dropped. It lies about a mile south of the Herdy Neb but has been covered by blown sand of late years.' Readers may not be surprised to learn that Herdy Neb, the only place in Cumbria with which the Devil is said to be associated, is just down the coast from Sellafield.

Irish Government Policy
According to the Irish government, Sellafield 'represents a potentially serious threat to Ireland's public health and environment as well as to our vital commercial interests, such as fishing, agriculture and tourism'. The Irish government would like to bring about the cessation of all activities at Sellafield. Its concerns relate, in particular, to the safety standards and safety management at the site; the storage on-site in liquid form of high-level radioactive waste; the continued reprocessing of spent nuclear fuel at the site; the transportation of nuclear fuels to and from the plant; the continued operation of the old Magnox reactors beyond their original design life; the discharge of radioactive materials into the atmosphere and into the Irish Sea; the risk of a catastrophic accident; the Mixed Oxide, or MOX, fuel fabrication plant.

The Irish government has repeatedly made known its concerns to the government of the United Kingdom. It has also highlighted these worries in international arenas such as the European Union, the International Atomic Energy Agency and the OECD, and at meetings held under the umbrella of the OSPAR Convention relating to marine pollution. Irish *taoisigh* have also personally expressed to British prime ministers the concerns of the Irish government about Sellafield and their opposition to the plant.

The present Irish government was elected in 2002 and consists of a coalition between the country's largest party, Fianna Fáil, and the small Progressive Democrats party. During the election campaign, these parties explicitly expressed their hostility to Sellafield. Fianna Fáil stated in its manifesto that it regarded the continued existence of Sellafield as 'an unacceptable threat to Ireland'. It added that,

> We believe that Sellafield should be closed in view of the devastating impact that a serious accident or terrorist attack could have on all aspects of life in Ireland, the ongoing risks from pollution of the marine environment, its poor safety record, and its lack of transparency and veracity with regard to its accidents as well as its normal operations.

Previously, in November 2001, the members of the Fianna Fáil parliamentary party had placed a full-page notice in the London *Times* to 'call on the British Government to reverse their decision on MOX and to shut the Sellafield complex'.

For their part, during the election campaign of 2002, the Progressive Democrats promised that they would

> vigorously pursue every option at international level, whether legal, political or diplomatic, to bring about the closure of the Sellafield reprocessing facilities. This will be done at international legal tribunals, at EU level and by forming an alliance to this end with like-minded countries such as the Scandinavian states.

The Irish government is acutely aware that some other governments, international organisations and various stake-holders see concerns about global warming, climate change and the need for sustainable development in energy as an appropriate basis on which to promote nuclear energy as the solution to these concerns. However, on this particular issue, the position of the Irish government, today, is quite clear. It states that the government flatly rejects the view that nuclear energy is sustainable because of the continuing and unresolved concerns that nuclear technology presents in terms of safety and accident risk, problems about the transport of nuclear materials, the issues surrounding waste management, the needless reprocessing of nuclear materials and increased proliferation risks. These concerns, allied to the high capital costs associated with nuclear energy, make it totally unsustainable.

Accident Risks and Ireland

One prominent theme of the Irish government's campaign against Sellafield has been the fear of a major accident at the site. The events of 11 September 2001 have made this fear more acute. The Irish government points out that, 'The Sellafield plant is no stranger to accidents.' It notes that, as recently as 1999, an investigation by the UK's Nuclear Installations Inspectorate (NII) into safety procedures and standards at the plant was prompted by a series of incidents that year. The subsequent report published by the UK Inspectorate in February 2000, on the control and supervision of operations at Sellafield, was severely critical of safety management and safety culture at the plant and made no less than twenty-eight recommendations for improving safety. The Irish government became deeply dissatisfied at the slow rate of implementing changes at Sellafield in the aftermath of that report.

The Irish government also considers the long-term storage of high-level radioactive waste in liquid form at Sellafield to be a serious risk. This waste arises at least partly from the reprocessing of spent nuclear fuel at the THORP plant. It is stored in twenty-one tanks at Sellafield, awaiting vitrification. BNFL has been directed by the NII to reduce stocks of the liquid waste to a buffer level of 200 cubic metres by 2015, through the vitrification process. The view of the Irish government is that the vitrification process, whereby the provide valuable information on its pathway across the country and identify the type and quantities of radionuclides deposited'. The air over Ireland is also continuously sampled by the RPII. Rainwater is collected routinely at twelve locations across the country. The RPII states that following the passage of any radioactive cloud, it could be expected that the areas of the country with maximum deposition would be where rainfall had occurred while the cloud was overhead. A 'Gamma Dose Rate Monitoring System' also consists of twelve stations, at which the gamma dose rate is recorded every twenty minutes. Readings are automatically transmitted to and stored on a computer database at the institute. In the event of a high reading an alarm is activated and the institute's Duty Officer is notified immediately. In the event of an accident, an additional programme of measurements can be provided by the Irish army's Radiological Monitoring Section, Reserve Forces.

The Radiological Protection Institute of Ireland is currently participating in a project under what is known as the European Union's 'Community Initiative Interreg II C': 'The central objective of the project, Atlantic Hydrology Modelling and Radionuclide Tracer Network, is to network a number of groups with an interest in the patterns of water circulation and radionuclide distribution in the Irish Sea, Celtic Sea, Western Channel and the adjacent Atlantic.' In other words, the RPII is joining with other interested parties in keeping an eye on how pollution from nuclear plants such as Sellafield makes its way around the seas and oceans of western Europe. Generally, marine discharges from Sellafield move northwards out of the Irish Sea.

In Northern Ireland, the Industrial Pollution and Radiochemical Inspectorate, which is part of the Environment and Heritage Service (EHSNI) within the Department of the Environment, monitors the impact of nuclear discharges into the Irish Sea.

waste is processed into glass cylinders to make it safer and easier to manage, should be speeded up considerably.

Radioactive Dumping in the Irish Sea

Radioactive waste discharges from the BNFL reprocessing plant at Sellafield constitute the dominant source of radioactive contamination of the Irish marine environment. There is no doubt that discharges from Sellafield result in contamination of the Irish and British marine environment and exposure of the Irish and British populations. Successive Irish governments have called for the ending of such discharges. The main pathway for this exposure of the Irish public to radioactivity is through the consumption of seafood. While the doses received through seafood or through marine-based activities are relatively low, and are believed by officials not to constitute a significant health risk at present, 'the fact that such contamination of the sea should occur is totally objectionable and unacceptable', according to the Irish government.

The Irish government states that it is 'intent on bringing about a cessation of reprocessing activities at Sellafield and resulting radioactive discharges, and will continue to work closely with its like-minded partners in OSPAR, particularly the Nordic countries, in achieving this objective'.

Challenging MOX

The history and operations of the Sellafield MOX plant were considered in some detail in earlier chapters. The Irish government says that it was 'totally dismayed and angered at the decision by the United Kingdom government to give the go-ahead to the Sellafield MOX plant' in 2001. It adds that, 'At a time when one would have expected countries with nuclear installations to consider the very real threat to safety and security and to consolidate safety and security standards, the UK Government's decision to effectively expand operations at Sellafield is difficult to comprehend.'

The commissioning of the MOX plant, on 10 December 2001, was regarded by the Irish government as both unjustified and unnecessary. The government had, earlier, repeatedly conveyed its strong opposition to this project to British ministers and department officials, and made detailed submissions during five separate rounds of public consultation held by the British authorities and ministers since 1997. The Irish government states that,

> We see no justification, economic or otherwise, for this plant. The MOX Plant will effectively perpetuate nuclear reprocessing activities at Sellafield and add to the level of radioactive discharges to the marine environment. It will increase the volume of worldwide shipments of nuclear fuels with obvious additional volume of traffic through the Irish Sea and thus pose an unacceptable safety and security risk as well as the potential for a major accident or terrorist attack.

Apart altogether from its objections to the amount of economic data and information which was deleted from the United Kingdom's public consultation documents relating to the MOX plant, the Irish government states that it cannot accept an economic analysis of that plant which writes off capital costs of some £450 million sterling already injected into the project. Furthermore, the Irish government claims that, 'there must be serious question marks about the projected markets for MOX fuel'.

Radiological Protection Institute of Ireland

The tone of expressions of anxiety by the Irish government is sometimes quite different from the tone of reports from the official Radiological Protection Institute of Ireland (RPII). Established in 1992, the RPII is the national organisation responsible for matters pertaining to ionising radiation. It describes itself as a 'watchdog against radioactive pollution of the environment'. Its reports are frequently cited by BNFL to allay fears about the possible effects of discharges from Sellafield on the Irish environment. They may be read on its website at http://www.rpii.ie.

In addition to having special responsibilities in the event of an emergency, the RPII concerns itself generally with hazards to health associated with radioactive contamination. The institute pays particular attention to monitoring the eastern coast of Ireland where, it states, 'the discharges into the Irish Sea from the nuclear reprocessing plant at Sellafield result in enhanced levels of radioactivity in the marine environment'. Periodic reports of such monitoring are published by the institute. In one report, in September 2000, the RPII noted that, 'Since 1994 the commissioning and operation of new facilities at Sellafield have resulted in an increase in the discharges of technetium-99 to the Irish Sea.' However, the activity concentration of caesium-137, which the RPII considers to be of greater radiological significance than technetium-99, was said to have stabilised. The RPII stated that,

> Clearly, continued exposure due to an installation from which Ireland derives no benefit is both undesirable and unnecessary and should be discontinued as soon as possible. However, the doses received through the consumption of seafood, walking on beaches or any other marine-based activity are low and do not constitute a significant health risk.

Samples of seawater, seaweed, sediments, fish and shellfish are regularly collected and measured by the RPII for a range of radionuclides. Across Ireland, samples of air, rainwater and drinking water are also tested for radioactivity. As a result of fallout over Ireland from the Chernobyl nuclear accident, elevated levels of caesium-137 are still found in mountain sheep in certain areas during the months of summer grazing. Flocks grazing in upland areas are regularly monitored by the RPII and live sheep are tested at abattoirs prior to slaughter. RPII measurement stations are distributed 'such that if a radioactive cloud passes over Ireland, these monitoring systems will immediately alert the institute to its arrival,

Postcards From Ireland

People in Ireland clearly feel strongly about Sellafield. And to demonstrate the point, in April 2002, an estimated 1.1 million special postcards were sent to England to demand closure of the Cumbrian site. 'Tony, look me in the eye and tell me I'm safe', read the cards directed to Prime Minister Tony Blair. Other cards went to Prince Charles and to Norman Askew, chief executive of BNFL. The postcard campaign was co-ordinated by Ali Hewson and other members of an organisation called 'Shut Sellafield' (http://www.shutsellafield.com/). Hewson is married to rock star Bono of U2 and lives in Dublin. The Irish postal service (An Post) distributed 1.3 million of the cards free to Irish households.

The postcards to Prince Charles displayed a radioactive shamrock, while those to the BNFL chief executive showed human lips and invited him to tell Irish people the truth about Sellafield. In response, a spokesperson for BNFL stated that 'there is no scientific evidence that our operations affect the health of Irish people'. Prince Charles declined to comment on the anti-Sellafield campaign, because it is a political matter. That the campaign reflected widespread public opinion in Ireland may be seen from the fact that the Irish postal service, An Post, so enthusiastically endorsed it. 'As the national postal authority An Post is delighted to have been able to serve the wishes of the Irish people in making clear their concerns about the dangers presented to public safety by the Sellafield nuclear reprocessing plant', said Liam Sheehan, general manager of Letter Post Sales & Marketing. 'Our final tallies indicate that on Thursday we will be despatching in excess of 1.1 million postcards to the three addresses in the UK. This figure is a major endorsement of the effectiveness of the post as a significant campaign vehicle', he added.

On the sixteenth anniversary of the Chernobyl nuclear disaster, Hewson personally delivered one giant card to Downing Street, bearing that same demand, 'Tony, look me in the eye and tell me I'm safe.' She pointed out that many millions of people in Britain live as close or closer to Sellafield as do the people of Ireland, suggesting that they should be as worried about Sellafield as are the Irish.

Not Getting Through?

The government of the United Kingdom was evidently underwhelmed by the postal protest. In a statement on 26 April 2002, the United Kingdom minister for energy, Brian Wilson, said that he recognised the genuine concerns of many Irish people and, where they had a rational basis, there was an obligation to respond to them. But, he continued, 'It does seem a little strange that millions of postcards have been printed, bearing the words "shut down Sellafield", when according to Ms Ali Hewson, speaking on BBC TV: "We know we can't close it down".' Wilson continued, 'It sometimes seems to be a case of denigrating Sellafield in a generalised and emotive sort of way, while ignoring the evidence produced from reputable scientific sources – most significantly, such sources within Ireland itself.'

In the House of Commons, on 1 May 2002, Tony Blair was asked by John Hume if he would confirm

> that he had received the largest amount of correspondence from individuals on one subject that any Prime Minister has ever received, given that more than one million cards have been received from citizens across Ireland expressing their deep concern about the safety of the Sellafield nuclear plant? Is he prepared to take the necessary steps to remove those concerns?

Blair replied to the representative from Northern Ireland that,

> I am aware of the concern that has been expressed. However, the Sellafield plant and any other plants in this country are subject to the strictest national and

international standards. Those standards are regularly reviewed. The plants are regularly inspected and none of those inspections has ever found a problem, such as the problem alleged in the press and by other political parties. Of course we take the concerns seriously, but there is a proper procedure and it would be wrong to close down nuclear facilities or start putting large numbers of people out of work without sufficient evidence from the relevant bodies to back it up.

Later, in July 2002, the number of cards sent to Britain was disputed. BNFL, which owns Sellafield, claimed to have received only about 150,000 postcards from the Shut Sellafield campaign, contrary to Irish estimates. In London, also, Prince Charles and Prime Minister Tony Blair acknowledged receiving far fewer Shut Sellafield postcards than An Post had said that it sent to them. However, a spokesperson for An Post was adamant that 'we completely and utterly stand over our figures'.

An Post repeated that approximately 250,000 (and not 'about 150,000') postcards had been sent by it from Ireland to BNFL's Norman Askew. It insisted that 276,000 had been despatched to Prince Charles, even though a spokeswoman for the prince said that Charles had received only about 200,000. An Post said that 584,000 cards had been sent to Tony Blair. However, in a reply to a parliamentary question on 10 July 2002, the British Prime Minister said merely that, 'Since January 2001 I have received over 187,000 letters and cards, including a very large number of pre-printed postcards as a result of a campaign organised in Ireland in April and May 2002.' Even if all of that 187,000 were pre-printed postcards, it still left a shortfall of up to 397,000 in the number of cards that Tony Blair was prepared to admit receiving. The use of the word 'over' by the prime minister was notably imprecise. Does it not matter to him exactly how many Irish protested by mail?

As well as the 1.3 million prepaid and pre-addressed cards given by An Post to Irish households, others went on sale in post offices and shops. An Post alone sold 250,000, at €1 each, and says it 'just about' covered its real costs in supporting the campaign to shut Sellafield.

Activists

Among those in Ireland who have struggled to raise public awareness about issues relating to Sellafield is a group of Louth residents who have been engaged in a long-running battle against BNFL. Also active is Earthwatch, or Friends of the Earth – Ireland (http://www.iol.ie/~foeeire). Based in Dublin, this small organisation has its work cut out trying to respond to a wide range of environmental issues. High on its list of concerns is Sellafield, although the organisation admits that it is difficult to make headway against the arguments and influence of the international nuclear energy lobby. Friends of the Earth, Ireland, regularly publishes *Earthwatch*, a periodical about issues relating to the environment in Ireland.

One organisation that has heightened the awareness in Ireland of the personal consequences of any nuclear accident is the Chernobyl Children's Project. It was created by Adi Roche and others to bring children from Chernobyl to Ireland for holidays. The publicity surrounding some of the visits of these children is a reminder of the harm that can be done to present and future generations by radiation.

People in Glasshouses

Although Irish people are very worried about the dangers posed by Sellafield, the Irish government itself was caught off-guard by the events of 11 September 2001. It emerged that, when it came to taking precautions against nuclear accidents, the Irish government had not got its own house in order. Following the attacks on New York and Washington, it transpired that the Irish contingency plan for international nuclear emergencies was very inadequate. The

Irish government was severely embarrassed when it became known that national stocks of iodine tablets, stored regionally and intended to be given to children and others in the event of a nuclear emergency at Sellafield or elsewhere, had not been kept up to date. Iodine can offset some of the effects of radiation.

That Interview

The Irish government's embarrassment at its preparations for a nuclear emergency being found wanting was compounded when the junior minister responsible for nuclear policy, Joe Jacob TD, went on a popular RTE radio programme after the attacks of 11 September and gave a performance that both amused and alarmed listeners. He was worked over by the programme presenter, Marian Finucane, and his interview was the object of much negative and sarcastic commentary in the media.

Martin's Mistake

And it was not only Joe Jacob who got into hot water. His more senior partner in government, the Minister for Health, Micheál Martin, attempted to reassure the public that there were ample stocks of iodine tablets, which would be distributed to the population to take in the event of a nuclear emergency. On Wednesday, 26 September 2001, Martin said that, 'They're stocked in the health boards in accordance with the 1992 plan and we have sufficient stock in all health board areas.' However, he was wrong. It soon turned out that some health boards had no usable stock. The minister was forced to retract his earlier assurance and now said that, 'It has emerged that some health boards have disposed of their stocks as they were past their best-before date and these boards believed they were ineffective.' He pleaded that he had given the earlier information in good faith. His department was said to be 'currently finalising the necessary arrangements for the purchase of new stocks of iodine tablets for pre-distribution in the event of a national nuclear emergency'.

A Big Plan

Attempting to repair some of the political damage done by his earlier radio interview, junior minister Joe Jacob assured Irish senators, on 10 October 2001, that,

> Concern about terrorist attacks on Sellafield or other nuclear plants has of course raised public concern about our preparedness to deal with the consequences of a nuclear incident . . . A National Emergency Plan for Nuclear Accidents is in place to ensure a rapid and effective response to accidents involving the release or potential release of radioactive substances into the environment which could give rise to radiation exposure of the public.

Jacob continued,

> The Plan is designed to cater for a major disaster at a nuclear installation in another country, which would result in radioactive contamination reaching Ireland. I should point out that the Plan is not designed to deal with a direct nuclear attack on Ireland but obviously a number of the arrangements and measures in the Plan would be relevant in such a scenario.

Once this plan is triggered, an Emergency Response Co-ordination Committee goes into immediate session in a central control room. The committee comprises representatives of key government departments and state agencies. Armed with information from the Radiological Protection Institute of Ireland (RPII) and other bodies (notably Met Éireann, because prevailing weather conditions will play a vital role), the committee will then decide on and co-ordinate the implementation of counter-measures and public safety advice. Information will be released throughout the course of the emergency by the RPII, using national radio and television and the Internet.

The minister advised senators,

Remember, we are concerned here with a nuclear accident not an attack, so normal methods of communications will remain operational. The public should stay tuned to bulletins which will provide information and best advice on a frequent basis. I want to stress that if there was a nuclear accident, for instance at Sellafield, and even if the wind is blowing in our direction, the contamination will take some period of time to arrive here. That period can be estimated and taken advantage of. We can identify the areas most likely to be affected; it would be unusual if the entire country was similarly affected. There will be time, for example, to get animals indoors and perhaps to take some limited steps to protect their feed in order to protect the food chain.

Jacob himself was not re-appointed a minister following the general election of May 2002, although his Fianna Fáil party was returned to power. Moreover, the nuclear safety brief moved from the Department of Public Enterprise to the Department of the Environment and Local Government, and its minister, Martin Cullen.

Keep On Taking the Tablets

In June 2002, eight months after it had emerged that the Irish government's plans for a nuclear emergency were in some disarray, each Irish household finally received an envelope of potassium iodate tablets through the post, from the Department of Health and Children. Just over 2.1 million tablet packets were distributed at a cost of €2.1 million. Marked 'National Emergency Plan for Nuclear Accidents', a warning in red letters told recipients to 'keep out of reach of children' and advised them that, 'This envelope contains a medicinal product and should only be opened by adults.' It seems appropriate that the tablets were manufactured at Runcorn in the United Kingdom, given that Britain is the location of the nuclear energy facility about which Irish people express the greatest anxiety. Inside each envelope were packed six tablets, with an expiry date of March 2005, and a leaflet. The leaflet explained that the tablets work by 'topping up' the thyroid gland with stable iodine in order to prevent it from accumulating any radioactive iodine that may have been released into the environment. It stated that the population groups most likely to benefit from taking the tablets are pregnant women, women who are breast-feeding, new-born infants and infants, children and adolescents up to the age of sixteen years: 'Priority should be given to these groups as the benefit to other population groups is limited.'

And for maximum benefit, advises the Department of Health and Children, 'the tablets should be taken before the radiation fallout reaches the area in which you live or work'. The leaflet does not mention school, leaving recipients to wonder what would happen to their children if Sellafield were to be hit by a jet on a weekday morning. Crushing and mixing the tablets with jam or honey, or dissolving them with milk or fruit juice, is said to be the best way to give them to babies and children. There is no reference to peanut butter or Coca-Cola. The tablets are not to be used after their expiry date, cautions the department, although some people might just chance doing so in the event of a nuclear catastrophe.

The envelopes with tablets were sent to households on the basis of the electoral register, with the result that some people living together received two envelopes, while large families received just one envelope.

In the same week as the envelopes were distributed, large advertisements were placed in national newspapers, headed 'Public Notice, NATIONAL PLANNING FOR NUCLEAR EMERGENCIES'. These gave advice on storing the tablets and noted that, 'In the event of a nuclear emergency, iodine tablets offer protection from radioactive iodine when associated with other principal counter-measures such as sheltering and avoiding the consumption of contaminated foodstuffs.'

The National Leaflet
Another informational leaflet distributed to Irish households
by the government, during the first half of 2002, and boldly
entitled 'National Planning for Nuclear Emergencies' is a
modest thing. It tells people how information and advice will
be given to the Irish public in the event of a nuclear
emergency, and adds that,

> As a result of such an emergency, radioactive
> substances released into the air could be carried in a
> manner similar to a plume of smoke, and could be
> deposited on the ground along the path of the plume.
> It would be several hours at least before any
> radioactivity reached Ireland. The amount of
> radioactivity in the plume would decrease with
> distance from the site of the emergency.

The public is assured that,

> There are early warning systems in place at national
> and international level.

This leaflet contains very limited practical advice on what
to do in an emergency, besides watching television and
listening to the radio, which is how the public spends a lot
of time anyhow. But certain counter-measures are specified to
protect people. These are, according to the leaflet, 'actions
designed to reduce the exposure of people to radiation in the
aftermath of a nuclear accident'. The principal counter-
measures are listed:

- Sheltering, i.e. remaining indoors (at home, in school
  or in the workplace) for some period in order to
  minimise exposure to higher radiation levels outdoors
  is a particularly effective and relatively simple counter-
  measure.
- Restrictions on consumption of contaminated foods.
  Where radioactive fallout has occurred, there is a risk
  of food supplies being contaminated and it may be

necessary to restrict their distribution and consumption.

- Agricultural measures with the aim of reducing the contamination of foodstuffs. These include appropriate sheltering and ongoing monitoring of livestock. Crops and other produce will be similarly monitored and restricted.

- Stable iodine being made available in tablet form (reduces uptake by the thyroid gland of radioactive iodine).

- Evacuation – it is not envisaged that an accident in a nuclear installation abroad would give rise to the need for evacuation of people in Ireland.

Deception and Kippers

One person decidedly unimpressed by Ireland's distribution of iodine tablets is Dr Rex Strong, a physicist employed as Sellafield's Head of Site Environmental Management. He says that taking iodine is only beneficial if there is a certain level of radioactive iodine in any discharge from a nuclear plant. He says that significant levels of iodine are unlikely to be released even in the event of a worst-case scenario at Sellafield. Strong describes the widespread distribution of tablets in Ireland as 'misleading' and 'a deception'. He adds that 'a pair of kippers has the same iodine content as these tablets'. In Cumbria, iodine is provided only to families living very close to the Sellafield site, because it is thought to be irrelevant in other cases. Strong says that when BNFL has followed up the provision of such tablets in England, the company has found that they are often lost by a family within a short time of distribution. He says that such measures may cause more harm than good by increasing the level of panic in a population at large if there is an accident. He suggests that greater damage was done by the official response to the nuclear accident at Three Mile Island in the United States than by the accident itself.

Hypocrisy?

Not everyone in Ireland feels the same way about Sellafield and Britain's other nuclear facilities as do the postcard senders. Three physicists wrote to the *Irish Times* during 2002, 'to put again the radiation from Sellafield into perspective'. Professor Walton of the National University of Ireland at Galway, Professor Mitchell of University College Dublin and Professor McAuley of Trinity College Dublin detected a note of hypocrisy in Irish opposition. They noted that it is said that Ireland gets no benefit from Sellafield and so should tolerate no radiation from it. 'We do, however, get some benefit in a number of ways', they wrote. They observed that Irish hospitals, researchers and industry use radioisotopes, some of which are produced by the British nuclear industry. Also, some Irish radioactive waste ('we have no nuclear repository') finds its way to Sellafield. The three physicists pointed out that Sellafield services some of the world's many nuclear power reactors and that these save greatly on oil and gas reserves, thus benefiting us all by keeping down greenhouse gas emissions. And they concluded with a query about possible future embarrassment for the Irish government: 'What is going to happen when the oil and gas run out? Northern Ireland is completing an electrical interconnector to Scotland and we have an interconnector to the North so, who knows, we might even find ourselves using British nuclear electricity' (*Irish Times*, 31 May 2002).

Vikings

During recent legal proceedings against Sellafield, the Irish government recalled that it has expressed to the United Kingdom its concerns about the impact of activities at Sellafield since the 1950s. Its concerns about reprocessing and the transportation of nuclear materials are shared by many other coastal states which also feel the impact on the marine environment of discharges from Sellafield. An informal alliance is being forged between Ireland and those parts of Europe from which the Vikings once descended to plunder our shores. The Irish noted that, on 31 October 2001, the Five Member states of the Nordic Council called on the United Kingdom to stop isotope pollution from Sellafield. Norway and other states have asked the United Kingdom to halt all radioactive discharges from the site and to close the THORP reprocessing plant. Norway has called for emissions from BNFL's reprocessing facilities to be processed inland and not to be discharged into the Irish Sea. The Norwegian Minister of the Environment has written to her United Kingdom counterpart expressing strong regret about the decision that the MOX plant was justified, on the grounds that, as the Norwegian minister puts it,

> the new MOX plant will strengthen the commercial basis for reprocessing activities at Sellafield and most likely expand the volume and prolong the life span of these activities as well as the discharges and risks they entail. There will also inevitably be more shipments of MOX fuel which represent a significant environmental and safety risk.

The shipment of nuclear fuels internationally has led to a growing number of objections from countries around the globe. For example, the ministers for foreign affairs of 'the Rio Group', meeting in Santiago, Chile, on 27 March 2001, were reported to have formally expressed their concern about the transit of radioactive materials and wastes along routes near their coasts, or along navigable watercourses of member countries, in view of the risks of damage involved and the harmful effects for the health of coastal populations and for the ecosystems of the marine and Antarctic environment.

In the course of exchanges during the legal proceedings relating to Sellafield that Ireland has taken against the United Kingdom, British lawyers dismissed such expressions of opinion about nuclear processing generally, or the transportation of nuclear materials by sea, as having 'no direct bearing, and in some cases no bearing at all, on the authorisation of the MOX plant at Sellafield'.

CHAPTER 18

Ireland Sues

The Irish government has taken two international legal actions against Sellafield. One of these cases has been launched under the OSPAR Convention, the other under the United Nations Convention on the Law of the Sea (UNCLOS). The Irish government has also considered invoking the Euratom Treaty in the European Court of Justice. The extent to which Ireland has committed itself to legal action on Sellafield is costly, and is unprecedented in Irish legal history.

Although Ireland itself once planned to adopt nuclear power, the Irish government has opposed Sellafield for over a decade. However, it cannot realistically bring a legal action based simply on its distaste for a particular British facility, or based on the negative opinions of most Irish citizens towards the nuclear industry. The Irish government acknowledges that nuclear power is widely used abroad, and that there is a framework of international agreements and organisations that facilitates the development of the nuclear industry.

Peaceful Promotion
Recognising that certain treaties provide for the peaceful promotion of nuclear power, Ireland's strategy has been to argue that peaceful promotion must involve appropriate measures for the transportation and disposal of nuclear fuels and for the operation and decommissioning of facilities,

among other measures. Development of the nuclear industry without such measures is a bird with one wing.

In this context, Ireland has sought to identify flaws in the procedures being adopted to continue operations at Sellafield, including the withholding by BNFL of important information relating to the MOX plant. It has also sought to challenge British assertions that the plant poses no serious threat to the environment. In the event of the Irish legal actions succeeding, the continuing operation of Sellafield as a reprocessing and fuel manufacturing facility may eventually not be viable. Even if the actions are not entirely successful, they may have the effect of rendering British Nuclear Fuels plc more accountable and ensuring the safer operation of its facilities at Sellafield and elsewhere.

The OSPAR legal action by Ireland began in June 2001, and relates to the fact that the United Kingdom, on grounds of commercial confidentiality, withheld pertinent information essential to assessing the 'economic justification' of the MOX plant. It is a prerequisite of the licensing of such a plant that it be economically justified to the satisfaction of the European Union. The other legal action, taken under UNCLOS, began also in the autumn of 2001. It relates to the fact that the Irish government considers that the United Kingdom has violated numerous provisions of this UN convention by withholding information on the MOX plant; and that the United Kingdom has failed to carry out a proper environmental impact assessment of the plant and of transports of radioactive materials. It also asserts that the United Kingdom, by permitting new discharges of radioactive materials into the Irish Sea, violates its obligations to protect the marine environment.

Interim Measures
The UNCLOS action, taken before the International Tribunal on the Law of the Sea (ITLOS), led to the establishment of an arbitration tribunal, under Annex VII of UNCLOS, to examine the merits and substance of Ireland's case. The

arbitration tribunal superseded the jurisdiction of the ITLOS Tribunal itself. Given the time expected to be taken in exchanging written memorials and replies, and to hold oral hearings (expected in the first half of 2003), the arbitration tribunal will possibly not be in a position to deliver a judgment on the substantial issues before summer 2003.

Meanwhile, the Irish government requested ITLOS to order the immediate suspension of operations at the MOX plant, pending the outcome of that arbitration. In particular, it asked for the following orders:

- That the United Kingdom immediately suspend the authorisation of the MOX plant dated 3 October 2001, or alternatively take such other measures as are necessary to prevent with immediate effect the operation of the MOX plant.

- That the United Kingdom immediately ensure that there are no movements into or out of the waters over which it has sovereignty or exercises sovereign rights of any radioactive substances or materials or wastes which are associated with the operation of, or activities preparatory to the operation of, the MOX plant.

- That the United Kingdom ensure that no action of any kind is taken which might aggravate, extend or render more difficult the solution of the dispute submitted to the Annex VII tribunal (Ireland hereby agreeing itself to act so as not to aggravate, extend or render more difficult the solution of that dispute).

- That the United Kingdom ensure that no action is taken which might prejudice the rights of Ireland in respect of the carrying out of any decision on the merits that the Annex VII tribunal may render (Ireland likewise will take no action of that kind in relation to the United Kingdom).

The United Kingdom requested the ITLOS to reject Ireland's request for provisional measures and to order that Ireland bear the United Kingdom's costs in the proceedings.

According to article 290 of the UNCLOS Convention, ITLOS may prescribe such provisional measures if it considers provisional measures appropriate to 'preserve the respective rights of the parties to the dispute or to prevent serious harm to the marine environment', and if it considers that certain requirements have been met, namely that *prima facie* the arbitral tribunal which is to be constituted would have jurisdiction and that the urgency of the situation so requires. By an order of 13 November 2001, the dates of the hearing of Ireland's case for interim relief were set as 19 and 20 November 2001. Following the hearing on those days, a decision was made by ITLOS and an order issued on 3 December 2001.

That order shows that, in the circumstances of this case, ITLOS found that the urgency of the situation did not require the prescription of the provisional or interim measures as requested by Ireland, in the relatively short period before the constitution of the 'Annex VII' arbitral tribunal appointed to hear the full case. However, ITLOS considered that the duty to co-operate is a fundamental principle in the prevention of pollution of the marine environment under Part XII of UNCLOS and general international law and that rights arise therefrom which the Tribunal may consider appropriate to preserve under article 290 of the Convention. In the view of ITLOS, prudence and caution require that Ireland and the United Kingdom co-operate in exchanging information concerning risks or effects of the operation of the MOX plant and in devising ways to deal with them, as appropriate. For these reasons, ITLOS did prescribe certain provisional measures, pending a decision by the Annex VII arbitral tribunal. It ordered that,

> Ireland and the United Kingdom shall cooperate and shall, for this purpose, enter into consultations forthwith in order to: (a) exchange further information with regard to possible consequences for the Irish Sea arising out of the commissioning of the

MOX plant; (b) monitor risks or the effects of the operation of the MOX plant for the Irish Sea.

Thus, ITLOS recognised Ireland's legitimate interest in the protection of the marine environment by directing that the United Kingdom should immediately co-operate with the Irish government in respect of the exchange of specific information about MOX.

Since that decision of ITLOS, the two governments have separately submitted reports to ITLOS relating to their co-operation. In ordering co-operation, the tribunal appears to have rejected British arguments that the Irish had not adequately attempted to exchange views on matters of concern relating to Sellafield. Partly because of that judgement, consultations between the British and Irish have taken place on the implications of the commissioning of the MOX plant for production levels, discharges and the transportation of fuel. The information exchange took place on a confidential basis.

Another Irish Case?
In the event of Ireland also taking a case to the European Court of Justice, the adequacy or otherwise of the economic case for the MOX plant will be considered under Directive 80/836/Euratom (as amended in 1984) and Directive 96/29/Euratom. These directives are also relevant in certain ways to the other actions taken by Ireland. The second directive, adopted on 13 May 1996, lays down basic safety standards for the protection of the health of workers and of the general public against the dangers arising from ionizing radiation. Of particular relevance is Article 6, which requires that, among other things,

> 1. Member States shall ensure that all new classes or types of practice resulting in exposure to ionizing radiation are justified in advance of being first adopted or first approved by their economic, social or other benefits in relation to the health detriment they may cause [i.e. 'economic justification'].

2. Existing classes or types of practice may be reviewed as to justification whenever new and important evidence about their efficacy or consequences is acquired.

Other Litigants

When the United Kingdom announced, in 2001, that the operation of the Sellafield MOX plant was justified, Friends of the Earth and Greenpeace challenged that decision in the British courts. Their challenge was dismissed.

In addition to itself suing the United Kingdom, the Irish government has also provided some financial support for certain residents of County Louth who believe that they have been adversely affected by emissions from Sellafield and who took technical advice in connection with their own legal action against the British. An earlier promise by Fianna Fáil that it would fully fund such an action was not kept when in government, ostensibly on the basis of advice from the office of the Attorney General of Ireland that such assistance would be inappropriate. The funding that has been provided was earmarked for research and for other supports that are a prerequisite of any legal action.

'Close to Zero'

One way in which the Irish government is currently challenging Sellafield is, as was indicated above, by taking a legal action under the international 'Convention for the Protection of the Marine Environment'. This convention is known as 'OSPAR', because it has been signed and ratified by all of the States that were also contracting parties to the Oslo or Paris Conventions, and it replaces those Conventions. Its signatories include Ireland and the United Kingdom.

In July 1998, under the OSPAR Convention, States adopted a strategy with regard to radioactive substances which commits the contracting parties to achieve 'progressive and substantial reductions of discharges, emissions and losses of radioactive substances'. This has the ultimate aim that, by the year 2020, 'additional concentrations in the marine environment above historic levels, resulting from such discharges, emissions and losses, are close to zero'. The adoption of this strategy was a recognition of the concerns felt by a number of the contracting parties, notably Ireland and some of the Nordic countries, about the impact of such discharges on the marine environment.

Since the adoption of this strategy, Ireland has been actively involved at subsequent meetings of the OSPAR Commission with a view to giving added impetus to the implementation of the strategy. For example, at the meeting of the OSPAR Commission in June 2000, Ireland tabled a draft 'decision' effectively calling for an end to nuclear reprocessing activities at Sellafield. In the event, at that meeting, the OSPAR Commission adopted, by majority vote, a decision whereby current authorisations for discharges of radioactive materials to the marine environment would be reviewed as a matter of priority by the respective national authorities (with a view to implementing the non-reprocessing option). The United Kingdom and France, both involved in reprocessing activities, did not support this decision, but its adoption effectively sent a clear message to the UK and France as to the concerns felt by the majority of contracting parties about continued reprocessing activities.

In July 2002, the British government announced that major cuts will be made to United Kingdom discharges of radioactive materials by 2020: 'The government is to ensure progressive reduction of concentrations of radioactive substances in the marine environment. The UK *Strategy for Radioactive Discharges 2001–2020* sets out how the UK will implement the Oslo and Paris Commission (OSPAR) Radioactive Discharges Strategy.' The statement added that, the strategy is based on plant closures, introduction of new abatement techniques and rigorous application of 'best practicable means' for reducing discharges. Speaking in connection with this announcement, the United Kingdom's

Secretary of State for the Environment, Margaret Beckett, said that radioactive discharges had already fallen. She said, 'Total discharges of beta activity from the site at Sellafield have come down to less than 1 per cent of their peak levels in the 1970s . . . The government is determined to maintain this downward pressure and to meet the OSPAR objective for 2020.'

Truly Microscopic?
In late October 2002, during oral hearings on the continuing OSPAR application by Ireland, legal representatives for the United Kingdom reportedly stated that radioactive discharges from the MOX plant are 'truly microscopic' and alleged that Irish representatives had been causing 'unnecessary public alarm'. For his part, Ireland's Attorney General, Mr Rory Brady SC, spoke of the impact of discharges from Sellafield on the Irish Sea and said that he was not simply making 'a dry, arid request for information'.

While Ireland's OSPAR application related principally to the disclosure of particular British documents, and not directly to the closure of Sellafield itself, freer access to relevant information allows Ireland to assess more comprehensively than otherwise any threats that may be posed to the environment by a continuation of reprocessing in Cumbria. The question of access to information also goes to the heart of Britain's economic justification for the MOX plant, which is a specific issue that may yet form the basis of a future Irish application in the European Court of Justice.

Ireland's Case Against MOX
Ireland's interim application to ITLOS revealed a great deal about the Irish government's fears concerning Sellafield. It relates directly to the MOX plant, to the international movement of radioactive materials, and to the protection of the marine environment of the Irish Sea.

In written arguments submitted to ITLOS, the Irish pointed out that the decision by the United Kingdom to authorise the MOX plant at the Sellafield site will further intensify nuclear activities in the Irish Sea. They noted that the Sellafield site is some 112 miles from the Irish coast at its closest point (at Clogher Head). They added that,

> Ireland has a special concern for its marine environment, not least since a significant proportion of its economy relates to fisheries activities in the Irish Sea, including in close proximity to the Sellafield site and the areas in which international movements of plutonium and other radioactive substances would occur. Under the relevant EU legislation, Irish fishermen may and do fish within 6 miles of Sellafield.

Along the Irish coastline, southwards from Northern Ireland, lie around 'fifty significant communities, whether cities, towns or villages, comprising a regular coastal population of some 1.5 million people (out of a total population of 3.8 million), a level which increases significantly during holiday periods', the Irish stated. The Irish reminded ITLOS that,

> In the case of the Irish Sea – which is indisputably a 'semi-enclosed sea' – the dangers posed by the increasing levels of radiation are clear . . . the discharges from the Sellafield site have already led to a steady increase in levels of radiation. Even though discharges of certain radionuclides have stabilised, or even decreased, the levels of radioactivity have not diminished. This is due to the long life of some of these radionuclides, and also to the physical difficulties of dispersing those radioactive discharges into areas beyond the Irish Sea. Increased levels of radioactivity have been detected.

The dangers from Sellafield itself are exacerbated by the shipment to Sellafield of spent nuclear fuel for reprocessing:

> In case of a shipping accident [said the Irish] spent reactor fuel elements contained in heavy casks could

sink to the bottom of the ocean and eventually corrode, releasing high level radioactivity into the ocean. The effect of an accident, involving the loss of some or all of the cargo in and around Ireland, would be to seriously contaminate the ocean and probably also the land with highly radioactive materials. This could have devastating effects on fisheries and on human health and the environment.

The Irish pointed out that BNFL is responsible for most of the activities carried out at the Sellafield site, being engaged in a range of commercial nuclear activities, including the reprocessing of spent nuclear power reactor fuels and the production of MOX fuel:

> It is expected to operate as a commercial entity. Sellafield is currently not a military site and it is not engaged in military activities.

Violations

In its Statement of Claim, Ireland identified to ITLOS a number of provisions of UNCLOS which it considers to have been violated by the United Kingdom. These provisions, said the Irish, obliged the United Kingdom

> to cooperate with Ireland in taking measures to protect and preserve the Irish Sea . . . to carry out a prior environmental assessment of the effects on the environment of the MOX plant and of international movements of radioactive materials associated with the operation of the plant . . . to protect the marine environment of the Irish Sea, including by taking all necessary measures to prevent, reduce and control further radioactive pollution of the Irish Sea.

The Irish claimed that the process used at the new MOX plant is unique and 'constitutes an experiment with unacceptable risks for Ireland'.

The Irish recalled that the reprocessing of nuclear waste fuel and discharges had begun at Sellafield (then called Windscale) in the 1950s: 'In 1993 a MOX production facility – known as the MOX Demonstration Facility (MDF) – began producing small quantities (8 tonnes per annum) of Mixed Oxide (MOX) fuel for Light Water Reactors. In 1994 the Thermal Oxide Reprocessing Plant (THORP) began operating, reprocessing spent nuclear fuel elements from Advanced Gas-Cooled Reactors (AGRs) and Light Water Reactors (LWRs), separating plutonium and uranium from fission products. A second reprocessing facility – the B2O5 plant – reprocesses spent fuel from Magnox reactors at Sellafield.'

Having recited this brief history of reprocessing at Sellafield, the Irish complained that,

> The MOX plant which is the subject of this dispute is intended by BNFL to significantly increase MOX fuel production for use in Pressurised Water Reactors (PWR) and Boiling Water Reactors (BWR). It is intended to have a maximum output of 120 tonnes of heavy metal per year. No nuclear reactors in the United Kingdom currently use MOX and so at present all the MOX fuel produced at this facility will be exported.

The Irish described how the production and use of MOX fuel involves three stages with significant implications for the marine environment. First, spent reactor fuel elements, containing plutonium and unused uranium and fission products, are transported to Sellafield, mostly by sea. Second, the spent reactor fuel is reprocessed at THORP where uranium, plutonium and fission products are separated; the plutonium, in the form of plutonium oxide, is then mixed with uranium oxide at the MOX plant to make MOX pellets, which are then placed into new fuel rods. Third, rods are assembled into fuel assemblies for use in nuclear power reactors and the fuel assemblies are transported from

Sellafield, again mostly by sea. The Irish pointed out to ITLOS that 'routine (intended) and accidental discharges of artificial radionuclides into the Irish Sea from Sellafield have occurred since the early 1950s. These discharges increased significantly in the 1970s, resulting in pollution that directly affects Ireland, including its waters.'

'Most Radioactively Polluted'
The Irish informed the ITLOS tribunal that, 'There have been many independent scientific assessments of the state of the Irish Sea which have concluded that as a result of radioactive pollution from Sellafield, the Irish Sea is amongst the most radioactively polluted seas in the world.' As an example, the Irish cited the report on 'Possible Toxic Effects from the Nuclear Reprocessing Plants at Sellafield (UK) and Cap de la Hague (France)', which was commissioned by the European Parliament's Directory General for Research, under the auspices of its Panel on Scientific and Technological Office Assessment (STOA). This is the controversial 'WISE' report which was considered earlier. The report was prepared by ten independent experts and submitted to the European Parliament in August 2001, but has been contested in some respects by other scientists. Its general conclusions included findings that marine discharges at Sellafield have led to significant concentrations of radionuclides in foodstuffs, sediments and biota; that the deposition of plutonium within 20 kilometres of Sellafield attributable to aerial emissions has been estimated at 16-280 GBq (billion becquerels), which is two or three times the plutonium fallout from all atmospheric nuclear weapons testing; that it has been estimated that over 40,000 TBq (trillion becquerels) of caesium-137, 113,000 TBq of beta emitters and 1,600 TBq of alpha emitters have been discharged into the Irish Sea since the inception of reprocessing at Sellafield (which means that 'between 250 and 500 kilograms of plutonium from Sellafield is now absorbed on sediments on the bed of the Irish Sea'); that in the UK about 90 per cent of nuclide emissions and discharges

from the UK nuclear programme result from reprocessing activities at Sellafield.

The Irish government pointed out that, according to the 'WISE' report, the reprocessing of spent nuclear fuel at Sellafield and at La Hague 'leads to the largest man-made release of radioactivity into the environment anywhere in the world'. The government claimed that the report had confirmed that nuclear reprocessing at Sellafield generates large inventories of radioactive waste which give rise to a significant risk of unplanned releases of radioactive materials.

Liquid Hazard

In the opinion of the Irish government, the greatest hazard from Sellafield is posed by the storage of high-level radioactive waste (HLW) in liquid form. Referring to about 1,500 cubic metres of such waste currently being stored at Sellafield in 21 tanks, it said that,

> Ireland considers that the current state of knowledge makes it difficult to prepare accurate evaluations of risk arising from such storage. Nevertheless, as the STOA ['WISE'] Report indicates, the consequences for human health and environment of an accidental atmospheric release from the high-level radioactive waste tanks at Sellafield would be far greater than the consequences of the Chernobyl accident in April 1986.

It may be noted that a number of prominent scientists share the opinion of the Irish government that the consequences of an accident at Sellafield could be extremely serious. These scientists include Professor Richard L. Garwin, who has advised successive US administrations on atomic energy and who recently co-authored a study of nuclear power and reprocessing.

The Irish government argued that the impact on the marine environment of discharges from Sellafield is felt on the quality of the waters and on marine life:

Lobsters and seaweeds, in particular, are known to concentrate radio-isotopes. The radioactivity can also contaminate beaches which would have deleterious impacts for human health. Moreover, the mere threat of such contamination could have potentially serious impacts on tourism, which is a very significant part of Ireland's economy.

'Poor Record'

In its presentation to ITLOS, the Irish government claimed that,

> There is a poor record of safety and compliance with regulatory authorisations at Sellafield, with numerous examples of violations of regulatory authorisations that continue up to the present. In October 2001 it was reported that BNFL closed its two Sellafield reprocessing plants because it could not reduce production of liquid high-level radioactive waste (HLW) sufficiently to meet regulators' requirements. In November 2001 a press report indicated the continuing adverse consequences of the 1957 accident at Windscale (Sellafield), with the Nuclear Installations Inspectorate reportedly halting the decommissioning of the Windscale reactor which caught fire in 1957 after an Inspector 'lost confidence in the Atomic Energy Authority's ability to carry it out safely and legally'.

The Irish recalled the revelations, in September 1999, that safety checks for MOX fuel destined for overseas customers at the Sellafield facility had been inadequate. Specifically, they referred to the fact that certain data relating to MOX fuel production at the MDF (MOX Demonstration Facility) had been falsified. The Irish told ITLOS that,

> An investigation launched by the United Kingdom Nuclear Installations Inspectorate (NII) of the Health

and Safety Executive was highly critical of the running of the MDF plant and reported as follows: 'It is clear that various individuals were engaged in falsification of important records but a systematic failure allowed it to happen. It has not been possible to establish the motive for this falsification, but the poor ergonomic design of this part of the plant and the tedium of the job [measuring MOX pellets] seem to have been contributory factors. The lack of adequate supervision has provided the opportunity'.

Underlining certain phrases in its own typed submissions in the case, the Irish noted that this NII report had concluded that,

The events at MDF [the MOX Demonstration Facility] which have been revealed in the course of this investigation could not have occurred had there been a proper safety culture within this plant. It is clear that some process workers falsified records of the diameter of fuel pellets taken for QA sampling. One example of falsification has been found dating back to 1996. There can be no excuse for process workers not following procedures and deliberately falsifying records to avoid doing a tedious task. These people need to be identified and disciplined. However, the management on the plant allowed this to happen, and since it had been going on for over three years, must share responsibility.

'Worthless'
Moving on from evidence of mismanagement of the systems at Sellafield to its financial administration, the Irish noted that the government of the United Kingdom had recently indicated a desire to dispense with its ownership of BNFL and to subject it to privatisation:

In that context there has been considerable attention paid to BNFL's current financial situation. According to the *Financial Times* the company is 'in balance sheet terms, worthless', with liabilities exceeding assets. This is a matter of great concern, given the potential legal liability of BNFL for damage resulting from the operation of the MOX plant, international transports, or other activities at Sellafield.

The Irish proceeded to attack the process of authorisation of the MOX plant. They argued that the process was 'badly flawed' and inconsistent with the United Kingdom's obligations under UNCLOS, because

the impacts of the MOX plant on the marine environment have never been properly assessed; no account has been taken of the impacts of international movements of radioactive substances associated with the MOX plant; material information has been withheld from the public, including in Ireland; and the United Kingdom has failed in its duty to co-operate with Ireland.

The Irish recalled to ITLOS that the MOX authorisation process had begun in the early 1990s. The first stage of the authorisation was the preparation of an Environmental Statement, assessing the impacts of the MOX plant on the environment:

On the basis of the 1993 Environmental Statement, which Ireland considers to be inadequate, the United Kingdom authorities approved the construction – but not the operation – of the MOX plant. The 1993 Environmental Statement has never been updated or revisited, despite longstanding and regularly repeated requests from Ireland.

It is also a matter of concern that no assessment of discharges, and no 'economic justification' was required in respect of shipments into and out of the United Kingdom of radioactive materials associated with the operation of the

MOX plant: 'As far as Ireland is aware, shipments have never been subject to any environmental impact assessment requirement, and their impacts on the environment have never been assessed.'

Moreover, prior to the construction of the MOX plant, BNFL already held authorisations for the disposal of certain levels of gaseous and liquid radioactive waste from the Sellafield site. In November 1996, BNFL submitted an application to the United Kingdom authorities for *variations* to these authorisations, including variations in respect of proposed discharges from the MOX plant. On this application, the United Kingdom Environment Agency formed the view that the proposed gaseous, liquid and solid discharges from the new MOX facility fell within the *existing* Sellafield authorisations, so that no new licence was necessary.

Economic Justification

European Community Law, as was noted above, requires the United Kingdom to ensure that a facility such as the MOX plant is 'economically justified' before being authorised. This means that the economic benefits of the plant should be shown to be greater than its economic costs. In respect of these obligations, according to the Irish government, 'the handling of this aspect by the United Kingdom Government does bear directly upon this Request [to ITLOS]'.

Between April 1997 and August 2001, the United Kingdom authorities held five rounds of public consultations on the 'economic justification' of the MOX plant. The first four rounds of consultations were based on a report prepared by an independent consultancy ('the PA Report'), and the fifth round of consultation was based on a report prepared by another independent consultancy ('the ADL Report'). The Irish objected that,

> The versions of the PA Report and the ADL Report which were placed in public circulation were heavily

censored, and most of the material financial and quantitative information was removed. The United Kingdom cited the grounds of commercial confidentiality, as well as the need to excise information 'on the grounds that the publication of the information would cause unreasonable damage to . . . the economic case for the Sellafield MOX plant itself'. The information excluded from the published reports related *inter alia* [among other things] to the volume of plutonium and uranium oxides to be reprocessed, the operational life of the plant, and the volume of international transports of radioactive material, including plutonium, associated with the plant.

The Irish lawyers appearing before ITLOS recalled that Ireland had made submissions in each round of consultation relating to the disputed economic justification of the MOX plant, and on each occasion Ireland had asked to be provided with a complete copy of the relevant report:

On each occasion the United Kingdom refused to accede to Ireland's request. Between 1994 and June 2001 Ireland made numerous and repeated written requests to be provided with the relevant reports. However, no substantive response was received to any of Ireland's numerous submissions or requests for information.

The lawyers recalled that, in a letter of 23 December 1999, Ireland had drawn the United Kingdom's attention to the significant change in the circumstances in which the MOX plant was to be authorised, which necessitated a review of the authorisation. On 15 June 2001, following efforts to resolve the dispute concerning the failure to provide information, Ireland had initiated proceedings against the United Kingdom under the OSPAR Convention. Ireland's lawyers maintained that the refusal of the United Kingdom to make available a full copy of the relevant information (including information relating to production volumes, international transportation and the costs of impacts on the marine environment) violated the United Kingdom's obligations under Article 9 of the OSPAR Convention. So, Ireland had initiated an arbitration procedure to obtain information on, among other things, production volumes, international transportation and environmental costs, in order to put itself in a position to assess whether the authorisation and operation of the MOX plant is compatible with the United Kingdom's international obligations.

In June and August 2001, Ireland had asked the United Kingdom to confirm that it would not authorise the operation of the MOX plant pending the conclusion of the OSPAR proceedings. After nearly three months of silence, on 13 September 2001, the United Kingdom had declined to provide such a confirmation. At that point, Ireland understood that the United Kingdom was determined to press ahead with authorisation irrespective of Ireland's interests and rights. On 3 October 2001, the United Kingdom authorities adopted a decision that the MOX plant was economically justified, and that over the course of its life it would make a net operating profit of between £199 million and £216 million sterling. That decision cleared the way for the operation of the MOX plant. The Irish argued that

> the manufacture of MOX fuel at Sellafield involves significant risks for the Irish Sea. Such manufacture will inevitably lead to some discharges of radioactive substances into the marine environment, via direct discharges and through the atmosphere. Manufacture is also vulnerable to accidents; and the MOX plant can only serve to increase the attractiveness of subjecting Sellafield to terrorist attack.

Advanced Powder
It was also claimed by the Irish that the operation of the MOX plant involves particular and additional dangers for the

marine environment, in part because of the particular characteristics of the MOX fuel which distinguishes it from other fuels:

> First, the MOX plant is an automated plant relying extensively on a software-based system for control of the process. Second, the production process involves the use of an advanced powder technology requiring the mixing, micronising, pressing, sintering and grinding of two actinide oxides. Experience in other powder processing industries, such as the pharmaceutical industry, indicates that technologies which are dependent on powder technology are not very reliable, since small changes in parameters (such as humidity, binder concentration and particle size distribution) can affect the powder and result in changes such as poor mixing or powder jams. Third, problems associated with powder technologies are exacerbated when, as in MOX fuel pellet fabrication, small batches need to be produced and variable formulations are pelletised. Fourth, lapses in the quality of inspections carried out by BNFL (for example in relation to plutonium and uranium isotopic composition) may have extremely serious safety implications and may have consequences which are time consuming and costly to rectify. Fifth, although MOX ceramic melts at a temperature of about 1,800 degrees centigrade, surface oxidation occurs at the much lower temperature of about 250 degrees centigrade if the fuel is exposed to air; at relatively low temperatures exposed MOX pellets give off respirable-sized particles following relatively short exposure periods.

The Irish also raised with ITLOS their concerns about the consequences of terrorism at Sellafield. They complained that, 'the United Kingdom has not consulted with Ireland – or sought to consult with Ireland – on the response measures it is taking to prevent terrorist threats or to address the consequences of any terrorist attack which might lead to the release of radioactive substances into the environment'.

The general failure of the United Kingdom to consult and co-operate with Ireland having been underlined by the Irish, Ireland's lawyers made a strong case for such future co-operation. They pointed out in late November 2001, more than two months after the events of 11 September in the United States, that,

> To this day Ireland has not been notified by the United Kingdom as to the proposed start date for the operation of the MOX plant, of the number of years over which the plant is to operate, of the volume of plutonium and uranium oxides which are to be reprocessed into MOX pellets, or the number of international transports of spent nuclear fuel and of MOX fuel assemblies which will be entering the Irish Sea in close proximity to Ireland. The United Kingdom has not notified Ireland of any emergency response plans in relation to accidents at the MOX plant or in relation to international movements of radioactive materials associated with the plant. Further, following the events of 11 September 2001, the United Kingdom has not notified Ireland of any additional security measures that have been taken or are proposed in relation to the Sellafield site generally or the proposed MOX plant and international movements of radioactive materials associated therewith in particular.

The Irish were to succeed in persuading ITLOS that the United Kingdom had been remiss in respect of providing information, with the result that, at the end of the interim proceedings, the court would order closer co-operation in future and would demand periodic reports from both governments to ensure that such co-operation would take place in practice. It is too soon to say if that decision by ITLOS has been fully accepted in its spirit by the British.

Irish Anomaly

The British response to Irish legal manoeuvres has been, as it were, to ask 'where's the beef'? Lawyers for the United Kingdom have argued that Ireland is concentrating on alleged procedural and planning flaws because there is no evidence that the MOX plant itself is causing serious harm to the marine environment, or is likely to do so. 'Instead of adducing cogent evidence of a threat to the marine environment arising specifically from the operation of the MOX plant,' retorted the British at ITLOS, 'Ireland relies on general assertions of dangers arising in connection with the nuclear industry or nuclear reprocessing or the practice of transporting radioactive materials or plutonium by sea.' They added that, 'Even a cursory review of Ireland's arguments reveals their weakness.'

The British highlighted, in particular, an apparent anomaly between the *Statement of Case* in the Irish application for interim relief from ITLOS and the main Irish *Statement of Claim* in the substantial case sent for arbitration before the 'Annex VII' Tribunal:

> In particular, whereas in the *Statement of Case* there is an allegation that the manufacture of MOX fuel involves significant risks for the Irish Sea, there is no such allegation in the *Statement of Claim*. It is difficult to see this as a mere oversight. This is, after all, the allegation that would appear to be at the heart of Ireland's allegations of breach of UNCLOS. The United Kingdom submits [that] the absence of an allegation of harm in the *Statement of Claim* reflects the true position in terms of significant risks to the Irish Sea caused by the operation of the MOX plant. As already shown in Part II, and as is considered further below, there are no such risks.

Irish 'misconceptions'

In their reply, the British also recalled the opinion of the European Union in 1997, cited earlier, that in normal circumstances the MOX plant posed no serious threat to the environment. They strongly denied that there had been an unwillingness on their part to exchange views or to co-operate with the Irish: 'If Ireland had accepted the United Kingdom's offer to exchange views, it would as a minimum have been disabused of a series of misconceptions of fact on which it bases its case.' For example, instanced the British,

> Ireland complains that the United Kingdom failed to discuss in confidence measures taken to guard against security risks. Had Ireland agreed to exchange views, the United Kingdom would have learned, at least, what are Ireland's concerns; and would have been in a position to determine whether they could be met. To take yet another example, Ireland complains that the United Kingdom failed in its duty to co-operate by withholding from the public domain certain information considered as commercially confidential relating to frequency of shipments. This, too, is a matter on which useful discussions could take place, on a confidential basis.

The British drew their own conclusions from what they suspected to be Irish disingenuousness:

> It may be that Ireland took the view that it would not be possible through an exchange of views to achieve what appears to be its real objective, to halt all operations at the Sellafield site. That, of course, is the case. But that does not affect Ireland's obligation to enter into an exchange of views in relation to the dispute under UNCLOS that it has now brought.

'Infinitesimally small'
Lawyers for the United Kingdom noted that,

> Ireland contends that there is a risk of discharges from the MOX plant to the Irish Sea. It is, however, important to bear in mind that the MOX plant is

essentially a dry process. The process itself does not give rise to liquid radioactive discharges. It is possible to anticipate some liquid discharge from the plant resulting, for instance, from the use of water in washing floors and fuel assemblies. This water will absorb some ambient radioactivity. It will, however, be treated and, after monitoring, discharged into the Irish Sea. The radioactive content of such discharges would be infinitesimally small. The same is true of any discharges through the atmosphere. BNFL characterises the annual combined liquid and gaseous discharges from the MOX plant as giving rise to a radiation dose to the most exposed members of the public equivalent to a dose received during 2 seconds of a flight in a commercial aircraft at cruising altitude or about 9 seconds spent in Cornwall in south-west England (this being an area underlain by granite).

Contending that there were various Irish 'misapprehensions' of fact, the government of the United Kingdom argued before ITLOS that,

The complaint that Ireland has brought before the Annex VII Tribunal is essentially procedural in nature. It is said that an environmental statement has not been correctly drawn up, that the justification exercise has not been carried through correctly, that information has not been supplied to Ireland, and that the United Kingdom has failed to publish or provide to Ireland an assessment of terrorist threats. It is not said that the Environmental Statement is wrong. It is not said that there is a risk of significant harm to the Irish Sea arising from MOX operations. It is not said that there is a significant terrorist threat arising from MOX operations.

To provide some perspective [added the British lawyers] the combined annual doses to the most exposed members of the public (for gaseous and liquid discharges from the MOX plant) would be less than one-millionth of the annual dose that the average person receives from background radiation occurring naturally in the environment. Doses to the critical group in Ireland would be considerably lower. The radiological impact on the general public from the MOX plant during plutonium commissioning and ramp-up to full operations will be smaller still than that from normal operations.

A Table Turned

The British used the expertise of Ireland's own officials against the Irish government's lawyers:

> Indeed, by submissions dated 4 April 1997 and 16 March 1998, the Irish Government acknowledged that any discharges were 'likely to be small'. Its position was that, irrespective of the low level of discharges associated with the operation of the MOX plant, it was opposed to any expansion of the operations at Sellafield. This is so even though the Radiological Protection Institute of Ireland ('RPII') has itself confirmed that radiation doses to Irish people continue to fall each year and do not pose significant health risks. In the words of the Deputy Chief Executive of the RPII, 'the dose to a heavy consumer of fish and shellfish from the northwest of the Irish Sea was 1.6 micro-sieverts in 1996 and 1.4 micro-sieverts in 1997. These doses are less that 1 per cent of the average dose of 3,000 micro-sieverts received in a year from all sources of radiation.' He went on to emphasise that 'it is safe to continue eating fish and shellfish from the Irish Sea and enjoying the amenities of our seas'.

The government of the United Kingdom adduced further details in its favour to knock down the Irish case. They recalled that Ireland had alleged that the impacts of the MOX plant on the marine environment have never been assessed and that no account has been taken of international movements associated with the MOX plant: 'This is wrong. The United Kingdom has implemented (and even exceeded) the relevant European and international regulations in its consideration of these issues.' The British played up the fact that the Irish case was against the MOX plant and not against THORP, or Sellafield in general. This allowed lawyers for the United Kingdom to dismiss certain concerns about the transportation of plutonium, for example, because such transportation was not directly relevant to the operations of the MOX plant itself.

'Spectre of Danger'
The United Kingdom responded robustly to the Irish allegation that the MOX plant poses 'significant risks for the Irish Sea'. The British contended that the allegation was based on a misguided view of the risks associated with the production of MOX fuel: 'While it is correct that the production process involves the production of wastes in solid, liquid and gaseous form, the quantities involved are such that there could not possibly be any harm – serious or otherwise – to the marine environment of the Irish Sea.' They said that Ireland had not explained how a dose to the critical group of 0.002 µSv (two thousandths of a millionth of a sievert) per year in respect of *gaseous* discharges from the MOX plant, or 0.000003 µSv (three millionths of a millionth of a sievert) per year in relation to *liquid* discharges, could cause harm to the marine environment of the Irish Sea. Nor had Ireland suggested that these figures, which the British described as 'the tiniest of fractions of the legally authorised limit', were wrong.

Moreover, added the British, 'while it is correct as the Irish say that plutonium dioxide in powder form is highly toxic,' 'the purpose of the MOX plant is to convert that plutonium dioxide powder into a ceramic state (MOX fuel) and then deliver it to customers safely and in accordance with all applicable international and national standards'. It had been suggested by the Irish that the production processes of the MOX plant might not be reliable. Yet, said the British, 'the support for this is "experience in other powder processing industries, such as the pharmaceutical industry", which are not remotely comparable with a nuclear plant subject to the [sic] stringent safeguards and regulatory controls'. They claimed that the Irish had offered no explanation as to how the alleged difficulties could harm the marine environment of the Irish Sea.

The British protested to ITLOS that, 'Ireland raises a spectre of danger and threat, but this is in the face of precise information on radiological impact that has been in the public domain for over eight years. No scientific analysis, no scientific data, no scientific opinion is brought into play to support this spectre.'

Conclusion

Walking about in the sunshine outside Sellafield's THORP facility, I feel reassured. Here are thousands of employees who appear to suffer no ill-effects from toiling alongside radioactive fuel and fission products. Individuals pass me like employees anywhere, preoccupied with the task in hand and intent on some normal activity; even looking a little bored. Occasionally, one nods or smiles at the person who is so obviously a visitor. These employees have a vested interest in keeping Sellafield safe, because they work on the site and live with their families nearby.

There is evidence which suggests that, in the past, particular workers at Sellafield were subjected to doses of radioactivity which had a detrimental effect on their health and on the health of some of their children. Now, workers are better protected by stricter laws and better regulations than once applied.

BNFL runs a well-regulated business and, usually, runs it efficiently and safely. The company has been faulted from time to time for errors and oversights at Sellafield, but these mistakes have seldom had any immediate recorded effects beyond the Cumbrian site itself. Certain discharges into the Irish Sea are permitted, but the official Radiological Protection Institute of Ireland and various British bodies charged with measuring effects see no reason to believe that discharges, at their present levels, will cause any great harm to humans or any significant damage to the natural environment. For as long as operations continue to go according to plan, Sellafield should not cause any big problems in the near future.

However, things do not always go according to plan, as the events of 11 September 2001 so vividly demonstrated. An attack by terrorists or by an enemy in time of war, a freak earthquake or accident, or an act of some deranged employee, could lead to consequences that were never expected or foreseen. The lives, health or livelihood of millions of people in Britain and Ireland, and beyond, might then depend on the effectiveness of British emergency services and of BNFL's contingency plans. How well would the shut reactors of Calder Hall, or the storage tanks full of highly active liquid or the new plants of THORP and MOX fare in a real test of BNFL's 'defence in depth' principle?

Unfortunately, it is not possible to assess precisely the likely response of British authorities to such unlikely events. Matters relating to security are shrouded in knowing assertions by BNFL and by British officials that all eventualities have been considered, but that it would be best not to reveal the details of security arrangements. However, a perfectly reasonable reluctance to give out information that might be useful to the malicious may also enable inefficiencies or inadequacies in the emergency services and in contingency plans to go unnoticed, until it is too late. A cult of secrecy in the past facilitated the creation of some of the very worst aspects of Britain's nuclear waste legacy.

BNFL is adamant that there will never be a catastrophic release of radioactivity from Sellafield. If they are wrong, the escaping plume may cross Scotland or England rather than be swept over the sea to Ireland. The fate of cities such as Edinburgh or Manchester, Leeds or Dublin, could depend on weather patterns at the time. In any event, the consequences of being exposed to such a plume would be very serious. Tens of thousands, or even hundreds of thousands, of deaths could follow and enormous costs would be incurred, including the provision of long-term health care and certain compensation for loss of earnings. There might be panic among the public, and a breakdown in law and order in some areas. The possibility that a ship transporting radioactive material to or from Sellafield might sink in the Irish Sea is also a matter of concern.

It is not just the immediate consequences of a disaster that are awful to contemplate. Far into the future, the area around Chernobyl will be blighted by the nuclear accident that happened there. The long life of radioactivity fills people with horror when they contemplate its particular characteristics. That is why it is very worrying that so much highly dangerous and radioactive liquid waste has been allowed to accumulate at Sellafield. The failure to 'vitrify' all of this waste into a hard form is a reflection on the standards of the British authorities and of BNFL, and the slow pace at which such vitrification has proceeded is inexcusable.

If Irish people assume that the standards of care in relation to the nuclear industry in the United Kingdom are among the highest in the world, they could be right. They might feel even less secure living next to some of the other countries in which nuclear reactors are located than they feel living next to Britain. However, the fact remains that serious mistakes have been made at Sellafield, and even some safety data have been falsified in recent years. Indeed, nowhere in the world is it ever possible to find perfection. Sooner or later, something dreadful may happen at Sellafield. No matter how hard people work to prevent such an eventuality, they can never entirely eliminate the chances of a catastrophe.

Sellafield is a reprocessing plant which would be less dangerous if it ceased to reprocess fuel. It could cease to do so very quickly, although BNFL's clients might sue for breach of contract and win substantial damages. But what cannot be achieved quickly is the complete closure of Sellafield. While the doors of THORP and the MOX plant might be padlocked tomorrow, there would remain for years a large inventory of highly radioactive liquids and other wastes, as well as contaminated buildings. Even were BNFL to speed up the process of vitrifying those wastes, there is no long-term repository for them anywhere in Britain. Buildings would also remain a radioactive hazard until fully decommissioned, demolished and taken in pieces to a nuclear waste dump.

The government of the United Kingdom is committed to reducing discharges at Sellafield, to speeding up the process of vitrification and to finding long-term solutions for the management of its nuclear liabilities. However, it is unlikely to close Sellafield in the near future. It has spent billions on THORP alone, and the operational future of Sellafield appears to be assured in the medium term. It is even more likely to remain operational if there is a reversal of trends in relation to the building of new nuclear power plants in western Europe and if the British continue to be involved in the important work of cleaning up nuclear facilities overseas. The environmental, commercial and strategic reasons why Sellafield still makes a lot of sense in Downing Street were considered earlier.

In normal operational circumstances, Sellafield appears not to be as dangerous as some of its critics fear. While BNFL has engaged in undesirable practices and has polluted the marine environment, there is an official consensus in both Britain and Ireland that the consequences of its doing so at Sellafield have not been severe. The Irish Sea may be the most radioactive in the world but, with some exceptions, scientists appear to agree that the level of radioactivity within it is relatively harmless.

Swimmers in the Irish Sea are probably more likely to suffer immediate ill effects from fertilisers washed into it, or from sewage or rubbish entering coastal waters, than they are from Sellafield's radioactive discharges today. And members of the public should perhaps worry more about the various additives and colourants that they willingly absorb through food and drinks, and through products such as toothpaste and shampoo, than about the level of radioactivity they are absorbing in the normal course of events.

However, in this context, words such as 'standard' or 'normal' are significant, for it is the danger of a sudden collapse of standards and of normal safeguards that makes

Sellafield peculiarly dangerous. Besides, the cumulative effects of the depositing of radioactive waste in the sea and on the seabed or the soil must be taken into account for the sake of future generations. Not enough is known about those effects to be entirely dismissive about the fear of harm from such continuing discharges.

It is the potential dangers posed by a catastrophe at Sellafield, rather than its daily operations, which provide a strong argument against its continued operations. Those dangers have not been diminished by the commissioning of THORP and of the Sellafield MOX plant, which almost certainly make the complex a more attractive target for terrorists and increase the level of certain discharges and wastes.

So long as it is safe, Sellafield appears to be very safe. But if and whenever it becomes dangerous, it could become very, very dangerous indeed. Sleeping with a loaded revolver on your pillow will cause you no harm, unless there is a certain degree of pressure applied to the trigger. Those who defend the continuing operation of Sellafield admit that there is always some possibility of a disaster but argue that the risk of a disaster actually happening is so low that it is reasonable to discount it. They claim that, even if Sellafield were attacked by terrorists, there are adequate measures in place to counter such an attack or to control its consequences.

But how *does* one calculate the chances of someone deciding to hijack an aircraft and fly it at a particular nuclear plant? And is it not implausible, in light of the destruction of the Twin Towers of the World Trade Centre in a matter of minutes, to imagine that one can confidently predict the effects of sabotage or terrorism upon the sprawling facilities at Sellafield?

In the case of Sellafield, the dangers of a catastrophe may be somewhat remote but their potential consequences are so great that it seems to me that the continued operation of Sellafield as a nuclear reprocessing plant is not justifiable. In coming to this conclusion, I have taken into account the existence of an alternative to reprocessing, namely dry storage of spent fuel and the manufacture of new fuel from fresh uranium. The fact is that the United States, which is the largest economy in the world, has survived without reprocessing for over twenty years.

To conclude that Sellafield should be shut down is not to conclude that all nuclear power plants ought to be closed. It has, hopefully, become clear in the course of this book that the potential risks posed by Sellafield are far greater than the admittedly real and grave dangers posed by a serious accident at any normal nuclear power plant generating electricity for the national grid. But the question of whether or not one is also to oppose the continued operation of the entire nuclear industry raises all sorts of difficult questions about renewable energy and greenhouse gas emissions that need not concern us when looking at the problem of Sellafield specifically.

Whether or not Sellafield closes, there will remain a requirement for some long-term facility in which to store the most dangerous nuclear wastes in Britain. Now that the government of the United States has chosen Yucca Mountain, the government of the United Kingdom will have to make its choice. This is unavoidable, unless we wish to see such wastes kept where they are at present, mainly at Sellafield and mostly in conditions that are far more dangerous than those of any properly designed long-term storage facility.

It is the case that the immediate closure of the reprocessing facilities at Sellafield would neither solve Britain's long-term problem of storing such dangerous radioactive wastes, generally, nor reverse the current commitment of many governments to nuclear power. It would also mean that the government of the United Kingdom and BNFL had uselessly spent vast sums of money on the construction of THORP and the MOX plant, and they might face further losses as a result of legal actions for breach of contract taken by BNFL customers. But closure would, for good reasons, reduce the level of anxiety among those who live in the shadow of Sellafield on both sides of the Irish Sea.

For now, people will just have to keep their fingers crossed. They can hope that the dangers from Sellafield's daily activities will continue to be minimised by means of technical controls and official regulation, while praying that the British authorities can cope successfully if something goes seriously wrong.

Some Further Reading

M uch has been published about the nuclear industry in general. Some titles that are relevant to Sellafield are listed below.

Official Publications

[Waverly Report] *The future of the United Kingdom atomic energy project* (The Waverly Report). London, 1953.

[Penney Inquiry] *Report on the accident at Windscale no. 1 pile on 10 October 1957* (The Penney Inquiry). London, 1957. This was also published as Appendix XI of Arnold, *Windscale 1957* (see Arnold).

[Parker Report] *The Windscale Inquiry*. London, 1978.

HSE. *The leakage of radioactive liquor into the ground, British Nuclear Fuels Ltd, Windscale, 15 March 1979*. London, 1979.

Charlesworth, F. R., Gronow, W.S. and Kenny, A.W. [for the Health and Safety Executive]. *Windscale: the management of safety*. London, 1981.

HSE. *The contamination of the beach incident at British Nuclear Fuels Limited, Sellafield, November 1983*. London, 1984.

Ministry of Agriculture, Fisheries and Food. *BNFL, Sellafield: revision of authorisation to discharge liquid radioactive waste: technical appraisal of information provided by BNFL*. London, 1986.

HSE. *The work of HSE's Nuclear Installations Inspectorate*. London, 1995.

Environment Agency and HSE. *The management of accumulated radioactive liquid waste and sludges at BNFL Sellafield*. London, 1996.

EU. *Council Directive 96/29/Euratom, 13 May 1996*.

Nirex. *Technical report: Sellafield repository design concept*. 1998.

Department of Environment, Transport and the Regions and Nirex. *Radioactive wastes in the UK: summary of the 1998 inventory*. London, 1999.

EU. *On greenhouse gas emissions trading within the EU*. Green Paper, COM (2000) 87. Brussels, 2000.

EU. *Towards a European strategy for the security of energy supply*. Green Paper, COM (2000) 769. Brussels, 2000.

NII/HSE. *The storage of liquid high-level waste at BNFL, Sellafield: an updated review of safety*. London, 2000.

NII/HSE. *An investigation into the falsification of pellet diameter data in the MOX Demonstration Facility at the BNFL Sellafield site and the effect of this on the safety of MOX fuel in use*. London, 2000.

NII/HSE. *Team inspection of the control and supervision of operations at BNFL's Sellafield site*. London, 2000.

NII/HSE. *The storage of liquid high-level waste at BNFL, Sellafield: addendum to February 2000 report*. London, 2001.

['WISE' or 'STOA' report] European Parliament Directory General for Research, Scientific and Technological Option Assessment (STOA) Programme. *Possible toxic effects from the nuclear reprocessing plants at Sellafield (UK) and Cap de la Hague (France)*. August, 2001. For this, see, http://www.wise-paris.org/english/reports/STOAFinalStudyEN.pdf

Department for Environment, Food and Rural Affairs, Scottish Executive, The National Assembly for Wales. *Managing radioactive waste safely: proposals for developing a policy for managing solid radioactive waste in the UK*. London, September 2001. A discussion document.

Performance and Innovation Unit [UK Cabinet Office]. *The Energy Review*. London, February 2002.

Department for Environment, Food and Rural Affairs, DTI and Department for Transport, Local Government and the Regions. *Energy policy: key issues for consultation*. London, May 2002.

Office of Civil Nuclear Security. *The state of security in the civil nuclear industry and the effectiveness of security regulation*. London, June 2002. http://www.dti.gov.uk/nid/index.htm.

DTI. *Managing the nuclear legacy: a strategy for action*. White Paper. London, July 2002.

EU/European Commission. *MARINA II*. Update of the MARINA project on the radiological exposure of the European Community from radioactivity in North European marine waters (up to the mid-1980s). August 2002. http://europa.eu.int/comm/environment/radprot/.

Department of Public Enterprise. *National planning for nuclear emergencies*. Dublin, 2002. An associated leaflet sent to households is at http://www.irlgov.ie/tec/energy/nuclear/national.htm.

DTI. White Paper on Energy. London, 2003.

Articles, Books and Monographs

Arnold, Lorna. *Windscale 1957: anatomy of a nuclear accident*. Copyright of the United Kingdom Atomic Energy Authority, 1st ed. 1992, 2nd ed. 1995. Macmillan Press, London.

Bertell, Rosalie. *No immediate danger: prognosis for a radioactive earth*. London, 1985.

BNFL. *Beyond U-235: an introduction to British Nuclear Fuels plc*. Sellafield Visitors Centre, 1988.

BNFL. *Nuclear energy: don't be left in the dark*. 1988.

BNFL. *For today, for tomorrow: environment, health and safety report 2000–2001*.

Bolter, Harold. *Inside Sellafield*. London, 1996.

Busby, Chris. *Wings of death*. Aberystwyth, 1995.

Carroll, John and Kelly, Petra (eds). *A nuclear Ireland?* Papers presented at an energy symposium, May 1978. Dublin, undated.

Cassidy, Nick and Green, Patrick. *Sellafield: the contaminated legacy*. Friends of the Earth, London, 1993.

Coeytaux, X., Faid, Yacine B., Marignac, Y. and Schneider, M. 'Airliner crash at nuclear facilities: the Sellafield case.' Paris, 2001. See WISE website.

Dean, G., Nevin, N.C., Mikkelsen, M., Karadima, G., Petersen, M.B., Kelly, M. and O'Sullivan, J. 'Investigation of a cluster of children with Down's syndrome born to mothers who had attended a school in Dundalk, Ireland.' In *Occupational and Environmental Medicine*, 57, xii (2000), 793–804.

Flood, Michael. *The end of the nuclear dream: the UKAEA and its role in nuclear research and development*. Friends of the Earth, London, 1988.

Gardner, M.J., Hall, A.J., Snee, M.P., Downes, S., Powell, C.A. and Terrell, J.D. 'Methods and basic data of case-control study of leukaemia and lymphoma among young people near Sellafield nuclear plant in West Cumbria.' In *British Medical Journal*, 300 (1990), 429–434.

Garwin, Richard L. and Charpak, Georges. *Megawatts and megatons: a turning point in a nuclear age?* New York, 2001.

Garwin, Richard L. 'The Garwin archive.' At www.fas.org/rlg.

Gowing, Margaret (assisted by Lorna Arnold). *Independence and deterrence: Britain and atomic energy 1945–52*. London, 1974.

Gubbins, Bridget. *Power at bay*. (The Druridge Bay campaign). Tyne and Wear, 1997.

Haszeldine, R.S. and Smythe, D.K. (eds). *Radioactive waste disposal at Sellafield, UK: site selection, geological and engineering problems*. Glasgow, 1997.

Henderson, Harry. *Nuclear power*. California, 2000.

Hewitt, Geoffrey F. and Collier, John G. *Introduction to nuclear power*. 2nd ed., New York, 2000.

Hussey, Matthew and Craig, Carole. *Nuclear Ireland*. Dublin, 1978.

Institute of Epidemiology, University of Leeds. 'The United Kingdom Childhood Cancer Study of exposure to domestic sources of ionising radiation: 2: gamma radiation.' In *British Journal of Cancer*, 86 (2002), 1721–31.

Lowry, David. 'A question of insecurity: the politics of assessing accident risks at Sellafield and La Hague – the story of the STOA ['WISE'] report'. Paper delivered at the fourth 4th UK and Irish local authorities standing conference on nuclear hazards, 21 March 2002. See http://nfznsc.gn.apc.org/ for this.

Mannix and Whellan. *History, gazetteer and directory of Cumberland*. s.l., 1847.

Melchett, Peter. 'The need for a public enquiry over THORP: decisions in the balance.' The *Guardian*, London, 1993.

Patterson, Walter. *Nuclear power*. Pelican, Harmondsworth, 1976.

Patterson, Walter. *The plutonium business and the spread of the bomb*. San Francisco, 1984.

Patterson, Walter C. *Going critical: an unofficial history of British nuclear power*. London, 1985.

Perera, Judith. 'Nuclear power in Europe.' *Financial Times*, London, 1999.

Robinson, Marilynne. *Mother country: Britain, the welfare state, and nuclear pollution.* New York, 1989.

Rowling, Marjorie. *The folklore of the Lake District.* London, 1976.

Royal Irish Academy. *Making sense of Sellafield.* Synopsis of papers delivered at a closed conference in Dublin, 26 September 2002.

Sadnicki, Mike and MacKearon, Gordon. *Managing UK nuclear liabilities.* Brighton, 1997.

Sheehan, P.M.E. and Hillary, I.B. 'An unusual cluster of babies with Down's syndrome born to former pupils of an Irish boarding school.' In *British Medical Journal*, 287 (1983), 1428–29. But see also Dean *et al*.

Taylor, Peter. *The environmental impact of a projected uranium development in Co. Donegal.* Political Economy Research Group, Oxford, 1985.

Taylor, Peter. *Consequence analysis of a catastrophic failure of highly active liquid waste tanks serving the THORP and Magnox nuclear fuel reprocessing plants at Sellafield.* Manchester, 1994.

Turvey, Frank. 'Sellafield: an engineer's view.' In *The Engineer's Journal*, 57, vii (Sept. 2002), 14–16; and viii (Oct. 2002), 14–16.

Uranium Institute. *The global nuclear fuel market: supply and demand 1995–2015.* London, 1996.

Walker, J. Samuel. *Permissible dose: a history of radiation protection in the twentieth century.* California, 2000.

Walker, William. *Nuclear entrapment: THORP and the politics of commitment.* London, 1999.

Western, Rachel. *Out of their depth: the inadequacies of Nirex research on the safety of nuclear waste disposal.* Friends of the Earth, London, 1994.

Williams, David R. *What is safe? The risks of living in a nuclear age.* Cambridge, 1998.

Wilson, Brian. 'Sellafield', The *University Observer*, 15 January 2003

Wynne, Brian. *Rationality and ritual: the Windscale inquiry and nuclear decisions in Britain.* British Society for the History of Science, 1982.

Zonabend, Françoise. *The nuclear peninsula* [Cap de la Hague]. Cambridge, 1993.

Periodicals and Annual Reports

BNFL. Annual report.

FOE (Dublin). *Earthwatch: the Irish environmental magazine.*

IAEA. Annual report.

NSD/HSE. *Statement of nuclear incidents at nuclear installations* (quarterly)

NSD/HSE. Annual Research Index.

NSD/HSE. *Nuclear safety newsletter* (three times yearly). See http://www.hse.gov.uk/nsd/nsdhome.htm.

RPII. Radioactivity monitoring of the Irish marine environment (occasional reports).

Index